To Maggie,
 Sincere thank
enlightening drawings which
have added a lovely, expert
dimension to this book, and
helped greatly to get the
message over.

Gill

("Susan McBane")

Fine Riding

FINE RIDING

Blending Classical Riding and Equitation Science

Susan McBane

CRC Press is an imprint of the
Taylor & Francis Group, an **informa** business

AN A K PETERS BOOK

Front cover: Anne Wilson's Pura Raza Espagnol mare Mill Sigilosa (Secret). Photo by Black Cat Photography.

First edition published 2022
by CRC Press
6000 Broken Sound Parkway NW, Suite 300, Boca Raton, FL 33487–2742

and by CRC Press
2 Park Square, Milton Park, Abingdon, Oxon, OX14 4RN

ISBN: 978-0-367-63895-5 (hbk)
ISBN: 978-0-367-63894-8 (pbk)
ISBN: 978-1-003-12119-0 (ebk)

DOI: 10.1201/9781003121190

Typeset in Times
by Apex CoVantage, LLC

Contents

Foreword

Unchanged for millennia, horse behaviour interpreted through human expectations has historically been subject to fads and fiats. Its capabilities and nature were ignored, from docking tails and ears to 'make the horse grow stronger', to insisting that the human dominate in every situation.

Throughout, the horse retains characteristics that, if not acknowledged, result in vices and endangerment to horse and human. Its reflexes are faster than a human's. Its nerves are so close to its bodily surface that it can respond to the minutest stimuli. It is acutely aware of the human's subtlest emotional signals. The near-imperceptible nod or lean directed Clever Hans's seemingly astounding mathematical abilities.

The authentic horseman is aware of these strengths, guards against their abuse, and develops them into a cooperative, beneficial relationship. Through the centuries, these aspects of the horse's nature have been incorporated into classical riding—the art of developing a lightness of aid and response, a mutual commitment between horse and rider.

Classical riding came into its own from fourth century BC horseman Xenophon's time through to the cavalry of the early twentieth century. Cavalry needed horses which could be ridden with one hand, guided by legs or shifts of weight, supple and balanced with economical movement. If there was any modification, such as stirrups or the forward seat, which increased equine comfort and efficiency, it was quickly adopted.

The classicists' main 'rock' is the exquisite balance, security, adaptability and communicative qualities of its famous default seat, a controlled relaxation as our author describes it. It is equalled by the goal of riding any breed or type with 'reins of silk' or, as I have heard it described: reins of rotten cotton. Until two or three generations ago, this quality was emphasised but is now rare.

The plea to work with the horse's mind and anatomy is under attack by the pressure of competition and obscured by tradition. The discerning eye views with dismay the prize-winning exhibits of the modern show ring: the hollow backs, the tail swishing, the inflexible pasterns, the tongue lolling indicating pain in mouth, hip or hind legs, tails broken and 'set', shoulders atremble with strain and pain, all from executing moves beyond what the physique can accommodate.

Classical riding is not to be confused with competitive dressage even at the Grand Prix level. It is, to the tutored eye, a readily recognised *style*. There is no room for the crowd-pleasing, judge-awarding showiness and exaggerated movement so attractive to sponsors and the 'riders in the stands'.

Equitation Science affirms the effectiveness of classical training. Developed almost two decades ago by Australian equestrians and scientists, Paul McGreevy of the University of Sydney and Andrew McLean of Equitation Science International in Australia, it is the catalyst for the International Society for Equitation Science. (*See* the *For Your Information* section at the end of this book). It grew from established animal learning theory into specific evidence-based equine learning theory. Its aim is to train a horse so that it understands us rather than having to guess how to respond to our aids or defend itself because it does not comprehend.

Equitation Science offers a system of groundwork which virtually ensures that the horse is reliably responsive to the basic aids before being backed or mounted. It delineates a clear progressive outlook along the training path. Should problems arise, it offers analytical tools to resolve them or prevent them in the first place.

For those riders around the world who have immersed themselves in any aspect of Equitation Science as an aid to classical riding, the reports are of an experience that is truly humane and superbly effective. The rider–horse connectedness that results is gratifyingly exhilarating. In the words of trainer Vicki Hearne, by making our communication with our horses 'kinaesthetically legible', the bonuses are incalculable. Expect a happier horse, longer lived with fewer behavioural and physical problems, the work of the trainer made easier, and increased safety for both horse and horseman.

Equitation Science, by introducing the Why, directs us to the scientifically validated How. Equitation Science together with classical riding is today probably the most humane, effective and practical way of training and enjoying the horse, competitive or not. To avoid getting involved is to accede to the status quo. Our horses deserve better.

Sharon E. Cregier

Sharon Cregier, Ph.D., FIASH (Hon., Edin.) was a founding member and North American Coordinator of British psychologist Moyra Williams's Equine Behaviour Forum which promoted sympathetic horse handling and the then revolutionary bitless riding. As a member of the international Animal Transportation Association, she has won three international awards for her work on improving horse transport. Retired from teaching research skills at the University of Prince Edward Island, she is a supporter of the Technical Large Animal Emergency Rescue work and continues to pursue all facets of horse behaviour through membership in the Animal Behaviour Society and the Animal Welfare Science, Ethics and Law Veterinary Association. She is guest editor for the open access journal, Animals, *specialising in horse transport and behaviour and co-edited the third edition of Fraser's* The Behaviour and Welfare of the Horse *(2021 CABI).*

Preface

THE BEST OF THE OLD WITH THE BEST OF THE NEW

Of all the thousands of equestrian books on the market, there isn't another one with the specific theme of this one—that of blending the 'time proven' principles of classical riding with the new, scientifically proven ones of Equitation Science, which is based on equine learning theory.

There are many different ways and schools of thought about riding and training horses, and many people have their own loyalties to a particular system to the extent of not being willing to look into others, to see the good or bad in another system, let alone give it a try.

This book is an introduction to what I believe, from experience, to be the two most humane and effective ways of training, riding and managing horses today, intended to give readers a working knowledge of both and, consequently, the ability to blend them so that the result is greater than the sum of its parts.

Classical riding is often seen as an esoteric school of thought and practice, a bit apart from other methods and not to be tampered with. Apart from the fact that all horses and ponies, whatever their circumstances, can benefit from classical principles and techniques, there are actually several schools of thought which are regarded as classical—the French way as practised at the Saumur school but also a slightly different and older French way now practised at Versailles, the Viennese way upheld at the famous Spanish Riding School in Vienna, the Iberian way which many believe to be the purest type of classical riding and other methods stemming from classicism, such as *good* Western riding.

It has to be said that, surprisingly, even at some classical establishments, we can still sometimes see ways of going that are not in accord with true classical principles, such as the poll being too low (not the highest point of the outline), 'shortened' necks, horses overbent, frothing at the mouth, behind the vertical and others, described later in this book. These are all hallmarks of force and restraint. The important thing is to be able to recognise them and not let them pass in your assessment of what you are seeing.

HISTORY OF CLASSICAL RIDING

Classical riding is often thought to have been named after the classical period of history usually regarded as the ancient Greco-Roman world which extended from the eighth century BC to the sixth century AD. It is generally believed that this period of history, specifically in those two partly interwoven civilisations, saw a rise in the arts, education and society to a state of maturity, justice, realism, quality and erudition that had not been seen in previous times.

Artworks from earlier centuries often show horses with what is now recognised as a demeanour betraying pain and distress, and it is not surprising. The fearsome bits and spurs depicted are not the equipment of a modern classical horseman or woman. Artists and other people then took these attitudes to represent fire, pride and sensitivity in horses, but today we are learning better due to the increase in equine scientific research in recent decades into how horses really learn and what they are, and are not, physically and mentally capable of, also how they differ in these respects from us and other animals. There is, too, a definite if slow shift in attitudes among the general public towards more humane treatment in the use of animals.

So, the expression 'classical' in the way we think of it today does not necessarily depict an era of history when all was rosy in equestrianism. To most of us, it harks back a few centuries, starting about the sixteenth century, when manège riding and the movements it involved was regarded as as much of an art as a necessary accomplishment for war—a horse light in hand, agile, manoeuvrable, instantly obedient to his rider's aids and ideally with a commanding, proud appearance and bearing that befitted his rider's status of royalty, aristocracy or gentility. This type of horse today is mostly associated with Iberian breeds and their cousins, Lipizzaners, but the principles of *true* classical riding are applicable to all breeds of horse and pony.

In those earlier centuries, horsemanship was one of the arts taught to royalty, the aristocracy and the gentry as part of their general education and social upbringing. An attitude of kindness towards animals was a sign of education and 'quality' but largely not because they did not want to cause them pain or distress. It was believed generally then that animals had no feelings of pain or emotion, a widespread if not exclusive belief that lasted well into the twentieth century. Even now in the twenty-first century, some people still maintain that fish and reptiles, for instance, cannot feel pain.

The education needed for a man to handle a strong, spirited but unpredictable animal like a horse was highly regarded because it made him physically and psychologically fit to (a) control himself and (b) control and lead a body of men, principally in battle. The classical training also fitted the horse to be a reliable conveyance in war, but outside of that, in parades, hunting, displays and sports as well.

Modern classicism is an ethos in which a horse's well-being is genuinely paramount, translated into physical movement and training, treatment and management, and into our attitude towards horses.

What does this mean in practice? It means certainly riding, and many people of a classical persuasion will include driving and other roles in this, at any level and in any activity which genuinely puts the horse's well-being first. Yes, it means lightness and promptness in responding to the aids of seat, weight, legs, voice and hands, also the whip-tap as a guide; it means the process of strengthening, balancing and accustoming a horse mentally and physically to move in such a way that, mounted, in-hand or at liberty, he can perform as asked with the least effort and risk to himself while offering relative safety to his rider or handler; and it means not forcing a horse to adopt a particular posture or carry out a requested movement for which he has not been equipped by heredity or training. As mentioned earlier, it also means not causing a horse pain or distress except temporarily, such as in veterinary treatment, and only when unavoidable.

This way of training and managing horses is not so quick or convenient for us but is *so* much more satisfying and ethical in every other way.

TELLING THE GOOD FROM THE BAD

Anything popular is often taken up and manipulated by the unscrupulous for prestigious or monetary gain. After the international Classical Riding Club was founded in Great Britain by Sylvia Loch in 1995, classical riding experienced a general surge in popularity and there soon appeared on the horse scene many teachers and riders jumping on the bandwagon claiming to be 'classical'. In short, and sadly, many of these men and women were anything but classical in ethos.

People of such a persuasion still exist today, claiming to follow classical principles but, in practice, using harsh methods to dominate horses, force them into 'outlines' (body postures) so that they *look* classical to the uninitiated, using strong bits, spurs, training 'aids' and whips harshly. Undoubtedly, this can cause considerable pain and distress to their own and clients' horses, forcing them to perform various training exercises and movements which are, in fact, counterproductive, difficult, painful, stressful, injurious and frightening for the horses so abused.

EQUITATION SCIENCE

They, and many people still today, treated horses as though they thought and learnt like we do, which we have known for nearly two decades now, through the development of equine learning theory and the gradual uptake of Equitation Science, that they do not. This applies to other animals as well; animal learning theory in general has been around now for many decades but, so far as the horse world is concerned, has been extremely slowly taken up. Admittedly, it is hard to let go of long-held, comfortable and much-loved ideas and theories, but these days there is no excuse for not doing so. Human behaviour change is of increasing interest today but, at the time of writing this book, still in its infancy. (However, refer to the *For Your Information* section at the end of the book.)

Many people eager to learn about classical riding put themselves unknowingly in the hands of trainers with highly inappropriate methods and attitudes and, to some extent, could not be blamed for putting their trust in someone who had a reputation for winning prizes in competition and was also very good at 'talking up' their practices and opinions. In fact, there seem to be very few classically trained horses in the wide competitive scene, and Equitation Science is disappointingly but not unexpectedly not yet widely popular.

I hope the information given in this book will make clear what is real classical riding and what is not. I also hope very much that the relatively new Equitation Science (ES), developed from scientific animal learning theory, will enlighten readers on a very important point for equestrians and horses alike—how horses think and learn as opposed to how we have so far assumed they do these things as, indeed, we do with other animals, too.

ES makes life so much clearer for horses, the end result being fewer 'bad' (actually, defensive) behaviours, fewer stereotypies ('vices'), more security and calmness and, therefore, better health and contentment. From our point of view, we do have to learn new methods and particularly timing of our aids, or cues and signals as they are termed in ES. The classical seat and its adaptations for racing and jumping is still needed in ES but, I think, its importance could be emphasised a little more.

Welcome to a new synergy in equestrianism. Whether or not you are familiar with classical riding, ES or both, or even neither, I believe the journey of discovery and slotting them together will prove fascinating and hugely rewarding for you and your horse.

Susan McBane

Acknowledgements

Most readers understandably have the impression that the main person involved in bringing a book to life is the author. As a very experienced author, I can tell you that nothing could be further from the truth. A book may or may not stem from an idea of the author, but it can easily be the brainchild of someone at the publishing company, a friend of the author, an idea sparked by something he or she has read elsewhere or a plea for a book about whatever from a member of the public.

In the case of this book, the idea came to me from a conversation with a client who was setting up a fine dining business. 'Fine dining', I thought, 'Fine riding', I then thought. 'Sounds like a good title for a book'. A bit more cogitation, and the idea of a book about combining true classical riding with Equitation Science worked its way into my brain. And here it is. I had been teaching the two together for years, but the book would not have materialised without the encouragement and very considerable help of several other valued friends and colleagues.

When working for an academic, technical or scientific publisher like CRC Press, authors first present their book idea and its proposed contents to the publisher and, if they like it, the proposal has to be peer-reviewed by others in the same field of expertise. In other words, they tell the publisher what they think of the idea, and of the author, and whether they think he or she is capable of writing it and delivering the manuscript on time, plus various other comments. Gulp! This, along with the weeks or months of waiting for their reviews, can be a somewhat nerve-racking process for the prospective author. Four peer reviewers is usual, and my four are people I have known and regarded highly for decades as friends and colleagues, if somewhat erratically, with whom I have studied and done voluntary work. This sounds like a mutual admiration society, but I assure you it is not. They all have strong principles and would not hesitate to tell the publishers not to waste their time and money on me, if that was what they felt. Fortunately, they didn't, so I should like to thank unreservedly, in alphabetical order, Dr Francis Burton, Dr Sharon Cregier, Mrs Sharon Foor and Professor Paul McGreevy. Thank you all so very much. I couldn't have done it without you.

I have a very special friend, Pauline Finch, whose generosity seems unbounded but who has no faith whatsoever in my ability to meet my deadlines. As a punishment, I coerced her, with virtually no notice at all and a ridiculous deadline, into drawing three diagrams on Equitation Science techniques in Chapter 7 of this book. Thank you *very* much, Pauline. You made your deadline—and, incidentally, I made mine!

They say, quite rightly, that a picture is worth a thousand words, and I am *so* fortunate to have been able to persuade superb equestrian artist and classical rider, Maggie Raynor, to illustrate yet another of my books with her *beautiful* drawings. She brings to them the eye of not only a rider but a real horsewoman and horse lover, and her work graces every book she illustrates. She is a dream to work with, too.

The advantage of drawings is that you can indicate *exactly* what you want to show. Not so with photographs. Any photograph is but a split second in time, and it is extremely difficult to capture in that split second the vital point, position or posture you need to show. My wholehearted thanks, therefore, go to my friend Lesley Skipper of Black Cat Photography for her unending patience in trawling through her considerable library of photographs to fulfil my all-but-impossible requirements. It has been a major headache for both of us but, thanks to her, we've done it.

Photographs were also provided by my good classical-rider friend, Anne Wilson—and her mares Lucy and Secret got in on the act in some of them. Thank you so much for showing just what I wanted to show. Still on the photographic topic, my gratitude goes to classical rider, teacher and author *par excellence*, Sylvia Loch, for giving permission to use photographs of her and her gorgeous Lusitano stallion, Prazer. I could not have wished for better. Finally, I wish I could thank personally that *doyenne* of classical riding and horsemanship in general, Sylvia Stanier, for photographs, friendship, China tea and everything I learnt from her over the years, but she is no longer with us. Instead, I must simply say: 'Thanks for the memories', many of them hilarious.

Lesley, together with John Barnes and Neil Harvey, also provided time-consuming technological assistance to this techie-dumbo, without which I truly don't know what I should have done. We all have our talents and digital technology is not one of mine. I know I am not the only one, but it is not very gratifying to realise that you are well and truly left behind, with no gift for catching up. Thank you all most sincerely. You haven't heard the last from me. My sincere thanks also go to Umamaheswari Chelladurai and Marsha Hecht of the production team, and especially to Cynthia Harasty, my Copy Editor, for their unending patience, understanding and professionalism, which I appreciate more than I can say. Artist Jonathan Pennel stepped in at the last moment to produce the vital sequence of six drawings of a pony jumping free, in Chapter 10. This is a particularly important sequence and I am grateful to him for pulling out the stops to get it done so expertly, on time. Thank you all so very much.

Now, The Boss—my Editor at CRC Press, Alice Oven, and her assistant, Damanpreet Kaur. Editors are rarely appreciated or understood by general readers, and why would they be? They work doggedly behind the scenes, pacifying their own bosses, flattering their authors, trying to get their manuscripts out of them on time, most tactfully pointing out where improvements could be made or where something just can't be said, getting into the heart of every book they handle with skilful rearranging, rewording and reworking, and come out smiling at the other end. I don't know how they do it. I once had a go at editing other authors' books and had to give it up as I was at risk of going mad. Alice and Daman have been so *patient* with my frequent queries, replied most promptly to them and always with practical, constructive advice and solutions. Without them and others like them, certainly there would be no books and a life without books, to me, is unthinkable. To all of you, I cannot express how grateful I am for your efforts, understanding and kindness.

Perhaps my readers may have gained some insight into just what a major team effort goes into producing a book, and I hope that they find this one enjoyable, thought-provoking and, at times, entertaining.

Susan McBane

About the Author

Susan McBane is known worldwide as a long-established equestrian author, having written 45 books, contributed to and revised several others for publishers and written hundreds of magazine articles. She has an HNC in Equine Science and Management, is a Classical Riding Club listed trainer and Gold Award holder, and an Associate Member of the International Society for Equitation Science. (For information on both organisations, see *For Your Information* at the end of this book.)

In 1978, with Dr Moyra Williams, herself an author, clinical psychologist, sport-horse breeder and intrepid horsewoman, Susan founded the Equine Behaviour Study Circle, later re-named the Equine Behaviour Forum, editing its members' journal for 30 years. Sadly, the EBF ceased operation at the end of 2019 due to dwindling member participation.

Susan has edited two commercial magazines, self-publishing one of them, *EQUI*, and is currently Publishing Editor of *Tracking-up*, a voluntary, non-profit quarterly which she produces with three friends. She has taught classical riding for 20 years, for much of that time combining it with Equitation Science, the blending of the two being the subject of this book. She has also acted as an expert witness, consultant, peer reviewer, judge and speaker on equestrian topics.

1 Two Approaches With the Same Aim

There have been many ways of riding horses down the ages. Sometimes it is done because humans enjoy it, but also at some point, it was realised how strong and fast horses are. Those are two qualities which are extremely valuable to humans. Gradually, horses (somewhere in the Far East so far as we know), having been hunted for food as wild animals up to about 6,000 years ago, gradually became domesticated. We probably did this by first capturing young animals, or maybe injured ones, then touching, feeding and watering and gradually taming them.

At first, it seems that horses were used for draught purposes, drawing *travois*, sled-type transport and, as the wheel was invented about 3,500 BC, early wheeled carts and other vehicles: this appears to have been horses' main use for a couple of thousand years or so. It seems probable that people worked out how to load up a horse with goods of various sorts on his back or carried by his sides in panniers. Maybe children, sick or elderly people and pregnant women were hoisted up there for transport, and so riding began. Brave youngsters may have vaulted on and off horses, fenced into corrals to keep them handy: they would soon realise that the back of a horse was a great place to be, quite comfortable, with a convenient mane to hang onto for security, and there was no turning back.

CLASSICAL RIDING

HISTORY

The ancient art of classical riding and the modern science of equitation—what can they possibly have in common? Like any other kind of riding, they both exploit horses for our benefit and enjoyment, but the main aim of both is to manage, handle, train and ride horses using *humane* and *effective* methods. Those are two words you will come across often in this book.

Classical riding is often thought to have started in eastern Europe, Greece in particular. Many horse people have heard of the Greek cavalry commander and military leader, Xenophon (born 430 BC, died 354 BC) whose book *De re equestri* (*On Horsemanship*) is still in print and enjoyed today. He advocated humane principles and techniques as he understood them. However, bearing in mind that the ancient Greeks, the Romans and other earlier civilisations had no saddles or stirrups, so were rather dependent on keeping themselves on their horses by means of the reins and bit in the heat of battle and other pressing circumstances, their bit contact must often have been far from light and comfortable. Any horse person who has visited the British Museum and seen the Elgin marbles or even just pictures of them must be struck by the depiction of the horses' head demeanour and facial expressions, most of which

DOI: 10.1201/9781003121190-1

Figure 1.1 François Robichon de la Guérinière, teaching his invention—the shoulder-in
(PD-US).

show clear signs of pain and distress stemming from their mouths. They would have
done well to have taken their example from the ancient Numidian cavalry, who went
into battle not only with no saddles or stirrups but with no bridles, either. Nevertheless, we have good reason to thank Xenophon for his book, the humane and often
insightful principles which are still very relevant today. Horses don't really change.

The 'father' of modern classical riding was François Robichon de la Guérinière
(Figure 1.1), whose life spanned the seventeenth and eighteenth centuries. A renowned
riding master and author, whose book *École de Cavalerie* (*School of Horsemanship*)
is still revered today and forms the basis of the teaching at the centuries-old Spanish Riding School in Vienna, he also devised the essential exercise shoulder-in which

he called 'the alpha and omega of all exercises' for suppling and strengthening and brought in the 'soft' (bent) knee. You will note in older illustrations that the riders always have straight legs, which must have made for a good deal of awkwardness in the saddle and discomfort for both horse and rider.

MODERN CLASSICISTS

Probably the modern master of classical riding most highly regarded is the late Col. Alois Podhajsky, head of the Spanish Riding School for many years. Podhajsky was responsible for the long trek out of eastern Europe during World War II to save the school's precious Lipizzaner horses from being 'horse-napped' by the approaching Nazi forces, and/or being slaughtered for food—a fate which did befall some of them.

Another classical master who also became a legend in his own lifetime was Mestre Nuno Oliveira of Portugal. Younger than Podhajsky, he was delighted to be able to meet, ride with and learn from him in Vienna, combining his methods with the established classicism of Iberia on his return home.

My own view of Oliveira's riding is that he never ceased to study and learn, developing increasing finesse throughout his life. If you source videos of him online, look for the ones of him as an older man, or compare the two periods. His trained horses ultimately always demonstrated the proverbial 'self-carriage on the weight of the rein', horse's weight back a little on the hindquarters, with the lightened and lifted forehand. I believe, according to my own classical trainer of the 1980s, Capt. Dési Lorent, who was a longtime friend and student of Oliveira and used his methods exclusively, that they lived long, sound lives. From the videos I have seen of Oliveira, I think he became a better rider as he aged, more 'feeling' and sensitive.

PANIC? WHAT PANIC?

Oliveira was famed for his rock-steady seat on a horse. Dési told me that, one day, a young horse in for schooling was having his first session with the master when a loud noise outside the school sent him up in the air, bucking and cavorting around the school for several minutes. Oliveira never budged an inch, never, of course, hauled on the reins, never shouted at the horse but just remained upright, soothing the horse till he came back to earth. 'His face was as red as a tomato', Dési said, 'but he never moved and just carried on as though nothing had happened'.

I saw him give a clinic in England in the 1980s, where a discussion developed in question time about bit contact. Another highly revered teacher, English, argued that Oliveira was teaching too light a contact for most horses and that weight-of-the-rein could be seen as *no* contact, or something like that. Oliveira asked him to hold the microphone flex while he kept the mike. He pulled quite firmly on it and asked: 'Would you call that contact?' 'Yes' came the answer. Oliveira lessened the pull but maintained a definite contact, asking: 'And would you call that contact?' 'Yes' came the answer again. Oliveira then let the flex loop and swung it from side to side. 'Would you call that contact?' he asked. 'Can you feel it?' 'Well, yes' came the answer. Oliveira just nodded and went back to his place in front of the audience to take more questions.

Another associate of Oliveira's told me he accompanied him to the classical military academy at Saumur in France, where he taught fairly regularly. One by one, the soldiers came into the arena on their beautiful French Anglo-Arabe and Selle Français horses. They rode round and Oliveira said almost nothing to them, as was his way, just the occasional comment, but he watched them closely all the time. At the end of each lesson, he simply said: 'Try to use a little less hand'. Excellent advice.

A final Oliveira story. At the previously mentioned clinic, a good friend of his, the late Sylvia Stanier, herself a renowned classicist (in fact, I think there was almost nothing she couldn't do with a horse), was accompanying him in the arena. Oliveira was explaining the famous classical seat and using Lucy Jackson, who had been Head Girl with Dési, to demonstrate. Dési had generously sent her over to work for Oliveira when he was looking for an assistant. He announced: 'That is the perfect classical seat', waving at Lucy. 'I myself am *not* perfect. I have one major fault when I ride. Is anyone here brave enough to tell me what it is?' Absolute silence, although everyone knew. 'No one?' He smiled and turned to Sylvia. 'Miss Stanier, will *you* tell everyone what the fault is with my seat?', knowing full well that she would. Sylvia, who loved a bit of fun, was smiling broadly and had a familiar twinkle in her eye. 'Yes, Maestro. You bow your head and look down at your horse when you ride. I've told you about it many a time'. Sharp intake of breath from the audience! Oliveira laughed loudly. 'YES, Miss Stanier, I do! Everyone is allowed one fault and that is mine'.

TRANSFORMATIVE DEVELOPMENT

One ground-breaking innovator, if not known as a classical rider, was the Italian cavalry officer, equestrian, instructor and author Federico Caprilli. He lived in the latter half of the nineteenth century and early years of the twentieth and devised the forward jumping seat to great effect, transforming the lives of both horses and riders. Old photographs of Italian military riders descending near-vertical slopes on forward-going horses, leaning *forward*, really put your heart in your mouth. Prior to the advent of the forward jumping seat, riders were taught to ride and jump with an upright torso and just a slight forward inclination on take-off, and to lean *back* on landing, lower legs often forward, almost invariably jabbing their horses in the mouth plus presenting them with the daunting burden of a significant, unstable weight swinging around on their backs during jumping.

I appreciate that most drop fences across country still call for leaning back with legs forward, if the rider is to avoid going over the horse's head on landing. Nonetheless, not all riders do this; the best learn to slip their reins to avoid trauma to the horse's mouth and give him freedom to recover and get away. In practice, he does this much better with a free head than by the rider pulling it up or using the reins (same effect) to stay on. It might seem helpful, but it isn't really. The best way a rider can help his or her horse is to be in good balance and give him the freedom to get you both out of trouble—not always easy!

In my own country, the United Kingdom, the best known true classical rider today is Sylvia Loch (Figure 1.2). A teacher, author and founder of the international Classical Riding Club, the point arrived, as it does with so many of us, when Sylvia wanted to partially retire and the active role of the CRC ceased. However, it is still

Figure 1.2 Classical rider, teacher and author, Sylvia Loch, founder of the Classical Riding Club, riding her Lusitano stallion Prazer, on the weight of the rein.

present online with a massive archive of thousands of articles by Sylvia and other classical authors (including one by me!) that were published in the members' magazine. A real treasure trove of information, education, entertainment and possible attitude adjustment, it is worth regularly visiting this invaluable free site.

What Do I Mean by 'True' Classical Riding?

Classical riding has a general image of proud horses and riders, horses with arched necks and tails, and wearing double bridles, performing the various advanced movements and sometimes the airs above the ground. The riders are seen to sit upright and proud, still and stable, and largely able to stay in their classical seat position no matter what their horses do. This image is not mistaken but it is a superficial perception of what is really going on within this equestrian pair.

I have heard many people say they do not like classical riding because the riders look so stiff and rigid, not daring to move for fear of upsetting their horses, who are oversensitive. They do not like that the horses are wearing double bridles; they believe they are performing 'useless' movements, that the riders aren't very good because they just need to sit there while the well-trained horses perform like clockwork, that they'd never win a prize in the show ring or a dressage arena—and so on.

The truth is that the riders are not at all stiff (see Chapter 7): if they were, the horses' movements would have them off pretty quickly because they would be unable to mould to the saddle, moving with their horses and dissipating their energy in what can be pretty gymnastic movements. Actually, their classical riders influence the horses invisibly with their sensitivity, weight, position, relaxed body control and subtle, light aids. The horses may be wearing double bridles, but classical riding can be done in the simplest snaffles and also in bitless bridles—or no bridles at all.

The movements the fully trained horses perform, advanced and above the ground, mostly had a purpose in the days when horses were used in war but also as a vehicle for royalty, the aristocracy and the gentry to display themselves in parades and displays. Their ethos came to be that these practices showed the true nature of the rider in being able to produce such difficult but correct airs in an animal as sensitive as a horse, and who conducts himself so instinctively. Animal welfare was a highly beneficial by-product, centuries ago, of the need to exercise kindness, patience and self-restraint when dealing with a highly mettled and very strong, easily panicked prey animal.

So, true classical riding trains horses to go on light bit contacts, light leg aids, to respond to natural weight aids (from the horse's viewpoint, to keep himself in balance under his rider and avoid falling) and to regard the whip as nothing to be afraid of. Because the mounted horse can see the whip and link its movement with the subsequent feel on his body, riders correctly use it as an extension of their arm to press or tap lightly on the parts his or her hand cannot reach. Sounds just like Equitation Science.

WHAT TO LOOK FOR IN A CLASSICAL TEACHER OR TRAINER

If you go to a classical demonstration or to watch a classical horseman or horsewoman you have heard of, you can tell if he or she is *truly* of the classical ethos simply by looking at the horse, watching him closely and noting his responses to what the teacher is doing to or with him, and what the teacher is instructing the rider or handler to do. *The horse will always tell the awesome, or awful, truth about his riders or handlers and the methods they are using.*

Chapter 3 is devoted to equine emotions, about which a good deal of research is going on at the time of my writing this book. We all think we know and understand the signs of the feelings being expressed by a horse, but some misunderstandings have been uncovered recently by research.

For instance:

• Ears pressed hard back and down do not indicate that the horse is 'listening/paying attention to his rider' but that he is angry and probably in pain and distress.
• Ears flicking loosely, maybe alternately, back and forth does indicate that the horse is paying attention to wherever his ears are pointing, even for a moment.
• Ears loosely positioned to the sides can indicate tiredness and fatigue or relaxation but also illness or possibly depression, not merely that the horse is taking a nap.
• The upper eyelid in a pointed, 'tented' shape indicates anxiety, not merely interest and attention.

- Dull, sunken eyes can indicate illness or a form of probable depression called 'learned helplessness' in which the horse is miserable with his lot and has learnt that nothing he does can improve it.
- Nostrils wrinkled up and back indicate annoyance, distress or pain.
- Lips not meeting, showing the teeth and maybe the tongue, chomping the jaws, champing at the bit rather than softly and occasionally moving it in the mouth, can all indicate anxiety, distress and discomfort or pain in the mouth.
- Upper lip over the bottom one, and a pointed chin can mean worry and fear.
- Significant froth around the mouth, maybe even splashing on to the chest and legs, indicates mouth pain and/or considerable discomfort: it is absolutely not a case of 'the more froth the better'. A working horse's mouth should be just moist, not sloppy.
- Tense, hard muscles can mean fear, tension and pain.
- Thrashing or swishing tail in work, as opposed to a relaxed, passive swinging of the dock and hair, indicates anger and distress, although, of course, in appropriate weather they can indicate irritation from insects.
- Tail clamped firmly between the buttocks indicates wariness and fear.
- Almost all these signs can also indicate confusion—a very common situation in modern, conventional riding (see Chapter 5).

The kind of work the trainer/rider is asking of the horse also gives away whether his or her heart and head are in the right places. Work on small circles, for example, ridden or in-hand, a great favourite with some clinicians, is extremely stressful for horses, especially at fast speeds, because their spines, contrary to popular opinion, are not laterally and uniformly very flexible. Working on small circles will not 'supple them up' but can cause injury, not only to the spine. Horses cannot follow smoothly the line of a circle or curve exactly, and they must not be worked on circles less than 6m in diameter. Performing four 5m circles along the short, 20m side of a manège, which is the party-piece of one pseudo-classical clinician I have seen, is ignorant showing off by the rider and damaging to the horse.

The horse's neck and tail are obviously the most flexible parts of the spine, but between the withers and the root of the tail, the three points of flexibility are:

- The tenth thoracic vertebra (in the first part of the torso).
- The first lumbar vertebra (in the back).
- The third sacral vertebra (in the croup area).

If the trainer has the horse tied up and strapped down with gadgets, this is a bad sign. As you will learn later in this book, it is possible, more effective and safer to train a horse without any training aids or possibly even side reins. The exception would be an abused or 'difficult' horse (and why is he difficult?) being rehabilitated or retrained, when loose side-reins could be fitted to come into effect only when the horse throws his head around or raises it beyond its natural height. In such a case, I feel the side-reins should not be fastened to the bit but only to the side rings on the noseband of a well-fitting, stable, lungeing cavesson, to save the horse's mouth.

Some clinicians and trainers have tremendous egos (for which there is no place in the horse or wider animal world) and feel they have to be seen to be successful (to 'Win') with a horse, come what may. Similarly, some are still working on the domination principle which is now known to be abusive and ineffective with horses, who do not have a clear dominance hierarchy in their natural relationships. People like this often ride or work a horse hard beyond when they should stop, and forcibly, in an effort to be seen to be a powerful boss, particularly if they want to prove a point. Depending on the temperature in the area being used, horses should not be worked up into a sweat beyond a gentle dampness of their coat, should not froth at the mouth, should not display any of the signs of pain and distress given earlier and should not appear tired, frightened, confused or worried. All these things are well outside the ethics of both classical riding and ES.

Of course, any trainer who whips a horse to punish him or 'make him do' (as opposed to giving light taps for direction) should be stopped at once, no matter who they are or the scenario, and the horse taken away from him or her. Other no-nos include the following:

- Using a firm and particularly a relentless bit contact.
- Pulling in and shortening a horse's head and neck posture.
- Working the horse in an over-bent posture (with the poll not the highest point of his outline and the front of his face behind the vertical).
- Jabbing or sawing the bit to punish the horse or 'make him sensitive in the mouth'.
- Pulling the horse's head and neck alternately to left and right with the bit in a purported (but ineffective and inappropriate) effort to supple him and make him 'listen' to the bit.
- Sitting badly in the saddle.
- Spurring a horse beyond a very light touch (which is not necessary, anyway).

These are not the ways of a true classical rider or trainer. Such people can verge on cruelty and certainly bullying, but others seem afraid to report them or to stand up in public for the horse. At least complain to the owner of the premises and the organiser of the event and consider writing to your country's premier equestrian magazine, even if you mention no names. If it's a horse for whom you are responsible on the receiving end of such treatment, stop the session immediately and take whatever other action you consider appropriate.

Social Pressures

When we are operating in some way inside a well-established community, large or small, with well-preserved, maybe 'time-honoured' practices and principles, maybe with a strong leader or keen followers, it can be very difficult to be seen to be 'different' and start using other methods. But if you come across some convincing reason to change, doesn't your horse deserve that you do so? You want the best for him or her, and for yourself. The fact that you are reading a book like this shows that you are looking for something better and, as Albert Einstein said (I understand): 'To keep doing the same thing and expect a different result is a sign of madness'. Quite.

We need to get back to the principle of always putting the horse first which was the standard teaching and general attitude until about the middle, or a little later, of the twentieth century. By that time, competition had become more and more popular and affordable as the world picked itself up and dusted itself off after the Second World War. By the 1970s, competition was very firmly established in the horse world and has never looked back. But now, it seems to me, winning often comes first, not the horses' well-being. It can be hard to go against the grain, but it's great to have a clear conscience!

It can also be difficult to stay independent of what others tell us. Many people tell us what they want us to believe, often very persuasively. We need to judge such statements rather than just going along with them and decide, with an open, rational mind, whether what is said or written is *really* in the best interests of the horse. Being able to put ourselves in the position of the horse is a great help.

WHAT ABOUT THE REST?

Some other ways of riding (see Chapter 5) can produce good results but often depend on the individual interpretations and talents of riders and trainers. Among them, there is often no clear path to follow, even in classicism. This is explained by the well-known saying: 'Every horse is an individual and must be treated as such'. Unfortunately, this can give people *carte blanche* to do what they want with their horse, to take differing advice from all and sundry, changing their methods by the week, and ending up with a very confused horse or pony. I once had an instructor who used to change her methods after every demonstration, lecture or clinic she attended: needless to say, I didn't stay with her very long.

Conventionally trained teachers and riders can have their own ideas of what to do, their own theories of how to go on and their own perceptions (usually and naturally human-based) of what a horse is doing, why and whether he is being good or 'naughty' or even 'bloody-minded'. Most have no scientific education to help them with the latter, and often ideas are repeated down the generations from teacher to rider, becoming ingrained as 'law'. 'This is how it's done' seems to be the catch-all excuse when no one ever really knows why or queries the underlying rationale. Unfortunately, academic and professional qualifications do not always mean you can rely on the advice given by their holder. I suggest asking yourself how you would like it if you were a horse, remembering what kind of animal a horse is, and 'if in doubt, leave it out'.

One of the most common reasons given to trainers by a new client performing a movement or giving aids in dubious ways is: 'My instructor told me to do it'. When they are asked if they understand the reasons behind it, they usually do not, or give an inaccurate, illogical or meaningless reply.

The trouble is that most of the conclusions reached by conventional trainers and riders are based, naturally enough, on human thought processes and horses are often treated like 'good' or 'naughty' children. However, some horses are worse off than naughty children because their riders readily flog them for perceived misdemeanours. This practice can be seen by the unknowledgeable as being dominant and strong, but it teaches the horses nothing except to associate being ridden with considerable pain.

One owner—not one of my clients—at a livery yard I taught at years ago used to 'punish' her horse for not winning a rosette at a show by not feeding him or even haying him up that evening. Explaining to her the illogicality, futility and cruelty of this fell on deaf ears and a closed mind, so the other owners made sure someone was always left on the yard after she had left, to feed and hay-up the horse. Also, somebody always arrived early in the morning to take out the bowl and empty haynet before she arrived back.

WHY RIDE THE CLASSICAL WAY?

Only a very few generations ago, the type of riding we now call 'classical' was simply the right way to ride, although, of course, human nature being what it is, not every rider bothered to learn it. There are many misconceptions about it, the most common being that it is too esoteric for the 'ordinary' rider and only about 'fancy stuff' such as advanced airs (movements), showpieces of no value to everyday or competitive equestrians—and too difficult anyway. Certainly, some of the advanced airs are difficult and appear pointless to a modern audience except as a display medium. However, the classical ethos of kindness and effectiveness benefits all our interactions with horses, involving training, riding, care and management.

Classical riding's main advantages are:

1 Its insistence on a balanced 'seat' or way of sitting, the rider's posture in the saddle and of using the position and weight distribution of that seat to guide the horse, and
2 Applying the aids (signals or cues) in such a way that they do not distress the horse with excessive pressures or create confusion in his mind, which is at the root of much defensive 'misbehaviour' or 'shutting down' in horses worked in the conventional, modern way.

In this, too, classical riding and ES have much in common. Conversely, it seems obvious to me that modern riding in general has strayed considerably in recent decades from older ethics.

So then, we need a way of injecting more consistency into our training and riding, a clear way of training, handling and managing horses that can be relied upon, that anyone, young or old, can understand, follow and apply. A way of riding that can be used and adapted for every horse, is humane, effective and does put the horse first. And now we have one—Equitation Science.

EQUITATION SCIENCE

Equitation Science (ES) is based on animal and specifically equine learning theory, with its insistence on scientifically evidenced, humane methods of training and riding horses, and continues to develop worldwide due to studies and research rigorously carried out in colleges, universities and other establishments. Because of it, there is now ample proof of how horses *really* think, why they behave as they do in-hand, under saddle, alone in their stables or at liberty with their friends or 'preferred associates'. We can now handle, manage, train and ride our horses using a language we know they will understand because it accords with how their minds are known to operate, rather than the rather confusing and often unjust one we may have largely and innocently used in the past. Horses are extremely good at adapting to us, but often we do not return the favour.

I have often described ES as the icing on the classical cake. It is certainly the most important development in equestrianism since the forward seat because it aims totally to improve equine welfare by making the experience of being handled, trained and ridden by people understandable to the horses and, therefore, safer for horses, riders and handlers. Most equine 'bad behaviour', evasions and resistances are actually self-defensive tactics in the face of distress, confusion or fear because the horses do not understand what is happening to them. Such techniques as bucking and rearing (the two most dangerous), bolting, kicking (which horses can do in all directions), biting, shying, spinning and others can seriously injure people: it makes sense to train and treat horses in such a way that they do not feel the need to use these defences.

HISTORY AND ORGANISATION

More and more scientists working in various fields are carrying out research studies into ES and new conclusions and discoveries frequently arise. By 2007, it was felt by the scientific and academic communities that it was time to bring it all together by founding the International Society for Equitation Science (ISES) which has resulted in an organised, voluntary, scientific body that promotes and publicises research and the practical application of modern equine learning theory to horsemanship worldwide. The ISES holds major annual conferences around the world which aid the dissemination and discussion of equine scientific topics. More information is given at the end of this book in the *For Your Information* section.

Two of the most prominent 'Names' in ES today are Dr Andrew McLean (of Equitation Science International, based in Australia, see *For Your Information*) and Professor Paul McGreevy of the University of Sydney. Dr McLean also does truly sterling work for elephants in relation to training with animal learning theory, and appropriate care and management, through the non-profit organisation HELP (Human Elephant Learning Programs). Both Dr McLean and Prof. McGreevy actively promote ES, and books they have authored are given in *Recommended Reading*, again at the end of this book.

More and more trainers and horse-people are also taking up ES, to the benefit of themselves and their horses. However, it really needs to become wholly embraced by our equestrian teaching organisations and sport administrative bodies worldwide. Gradually, even if it takes a generation, I hope that ES based on equine learning theory will become imbued into the international horse world just as animal learning theory (from which equine learning theory was developed) has become almost universally accepted in the wider animal world.

I have been studying, using and teaching the principles of ES along with classical riding since I read Andrew McLean's introductory book on the subject—*The Truth About Horses* published by David & Charles in 2003—and find that the two blend together ideally, with excellent results and, as stated, both have the same aims of effective, humane principles and techniques.

Equitation Science is not just another system or method of horsemanship. Rather, it is a collection of principles that can be applied to any method, any horse or pony, in any job, that we can use to create best practice in their training, riding, handling and management, to enhance equine welfare and well-being.

OUR SOCIAL LICENCE TO OPERATE

Society as a whole is becoming increasingly aware that humans' treatment of animals is often not as humane as it could be—and is increasingly willing to make its feelings known. Various groups and individuals express their objections in different ways, from letters to magazines and newspapers, peaceful demonstrations, and strong representations on social media to violent protests which, however laudable their purpose, are often against the law and can result in damage to property and injury to people and, indeed, animals.

The point is, we in the horse world know that there are welfare and well-being issues in our milieu. Times change and practices change. People still use techniques in the work and management of horses that have been accepted for not just decades but generations. Due to rigorous research studies, we now know more than ever about equine biology and biomechanics, and the veterinary and physical therapy care of our horses has improved immeasurably as a result.

Thanks to ES, we are fast learning more and more about horses' mental, psychological and behavioural functioning. This enables us to change our riding, training and management techniques to better suit the type of animal we are dealing with and so improve the quality of horses' lives, their safety and our safety and enjoyment, too.

There is still a long way to go, but if we do not make the journey, consign some principles and practices to the past and embrace, individually but also via our teaching establishments and administrative organisations, more humane, scientifically and practically proven improvements in our dealings with horses, the general public will gradually see to it that we lose our Social Licence to Operate.

My experience of using authentic classical riding and Equitation Science together convinces me that this blend of the best of the old with the best of the new is the way forward. True classical riding stands the test of time. Equitation Science is standing the test of science. They have many principles, techniques and practices in common and I think the two together are unbeatable. Do give them a try.

IN A NUTSHELL

True classical riding and Equitation Science (ES) have a great deal in common, most importantly that they both aim to put the welfare of the horses first. They are easy to blend because ES is not a method or system on its own. It is a fairly recent collection of principles that can be applied to any method, any horse or pony, in any role, that we can use to create best practice in their training, riding, handling and management, to enhance equine welfare and well-being. Classical riding, on the other hand, has been around for thousands of years in various forms but not everyone who claims to be a classicist uses true classical principles.

We have covered the history of both classical riding and Equitation Science, and particularly clarified the difference between true classical principles and the techniques we see from some practitioners that are not in the best interests of the horse— in short anything that stresses out the horses—plus how to recognise and so avoid them and those who promote them. The need to return to putting the horse first has been stressed, plus the value of being able to put ourselves in the horse's place. Traditional and also conventional, modern riding are also covered, with advice to not keep doing what has always been done just because it has always been done.

Equitation Science has been explained as being based on proven scientific work aimed at communicating with horses in a way they can understand and respond to as we intend. It is a reliable and trustworthy way of training. It is the most important equestrian development since the advent of the forward jumping seat: furthermore, it makes horses calm and biddable because they understand what is happening to them and how to react.

Finally, we consider our 'Social License to Operate'—the level of acceptance the horse world needs from the general public, who are watching animal welfare issues ever more closely, for it to continue as a recreational, professional, commercial and domestic activity in the future.

2 What Are We Dealing With?

It has been said that of all the animals we have domesticated for our own requirements, horses have been the most useful to us.

They have a unique mixture of strength and speed, two qualities most obviously valuable to us. They can also jump (seemingly a legacy of their ancestors' forest beginnings where they would have had to jump over fallen trees and other obstacles), they are intelligent and biddable enough to train, they are sensitive and they are sociable not only with their own kind but also with other animals and humans. Despite being creatures of the wide open spaces more recently in their evolution, they adapt to being confined in stables although to a lesser extent than many of us believe, to travelling in moving vehicles, to having metal shoes hammered on to their feet, to wearing tack and clothing, having their teeth rasped and most other things humans do to and with them.

EVOLUTION

Their evolution gave horses various other qualities which we need to accommodate if we are going to have a mutually beneficial relationship with each other. They evolved as prey animals, latterly in open areas with few hiding places from predators. Some equidae still adopt the tactic from their forest-dwelling times of hiding behind anything available when they sense danger. This is possibly because hiding was more appropriate in a forest than trying to run away around and over trees, shrubs and ground littered with broken branches, tree trunks and dense undergrowth.

As the climate slowly changed and the earth's environment changed with it, plains appeared supporting new species of vegetation—the grasses. Evolution depends on the ability of genes to mutate: those very early, small, equine predecessors whose genes mutated to produce dentition and physical features more suited to grazing than browsing, and longer legs, necks and heads which enabled horses to reach the ground and gallop at high speeds, were more successful at surviving in this new environment than those who retained their forest-animal features.

Other animals were changing, too, and the type of predator open-living animals met with was largely different from those in the forest. For both predator and prey, speed became crucial to survival, so that the predators could run and catch prey and their prey, conversely, could run and escape. The skills of jinking and turning at speed as well as flat-out galloping were also developed and were at a premium.

DOI: 10.1201/9781003121190-2

MODERN HORSES' CHARACTERISTICS

SPEED

Modern feral horses living in open environments have been clocked variously as running at around 40mph/64kph, cheetahs at around 50mph/80kph and racing greyhounds between the two. Leopards retain their forest ancestors' inclination to live partially in trees, so an animal able to climb a tree may not be safe from a leopard.

Luckily for us, horses don't climb trees although I know one who has tried it, and tried to climb a ladder: he has also worked out the technique of rearing up, resting one fore hoof on the branch of an apple tree, bashing at apples with the other, or swinging it to and fro with both hooves, until some apples fall off. Unfortunately, they are cooking apples, so his owner has to clear them daily in autumn to prevent the horse from getting colic, but he still manages to get a share—all to himself, as he is one of those few equines who endanger the lives of others turned out with them.

Probably we all realise that once a horse has been really excited, or badly scared, he can remain alert and hyped up for the rest of the day because the hormones take some hours to return back to normal levels. Again, training correctly, especially

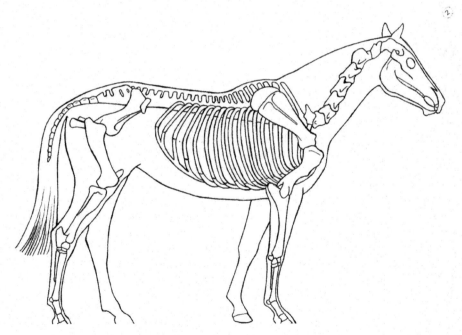

Figure 2.1 The horse is a herbivorous, running prey animal. His skeleton is superbly adapted for this lifestyle. The head on the end of a fairly long neck enables him to reach to the ground to graze grass and to reach up to shrubs and trees to browse leaves. The legs are long enough to enable him to gallop at speed away from most canine and feline predators in the open-space environment to which modern horses are adapted.

getting the horse to stand still or walk very slowly and carry his poll below his withers, can reduce this effect, and the Equitation Science techniques described in this book are far more precise and reliable than some conventional methods. Punishing a horse in any way in such a situation effectively compounds his fear, which would mean more trouble for ourselves and the horse in the future.

Speed is a horse's main, natural way of staying alive in the face of predation. His phenomenal ability at standing starts, sparked by his alertness, gives him a good chance of getting up to his top speed in a very few seconds and avoiding being caught. Horses, then, are obviously fastest when galloping on a straight line which is why we like racing them, but they must slow down somewhat when cornering, and their usual predators—canines and felines—are good at this.

Sharp turning and cornering put a great deal of stress on horses' limbs: they are not lightweight animals and their bones and joints have just the same basic composition and structure as the bones of a human or an elephant. In some forms of riding and competing, horses are expected to corner very sharply carrying not only their own weight but also that of a relatively heavy saddle and a rider. As we now use horses for our own leisure purposes, surely it is time that these sorts of activities purely for sport should be stopped.

JUMPING

The ability to jump has been retained in many equines from their forest-dwelling ancestors, who would have had to negotiate fallen tree trunks and other obstacles. Horses use their heads and necks to help them balance as human athletes use their arms and hands. Ask a benevolent friend to strap your arms to your sides at the elbows and wrists and then try to run and jump round an obstacle course, and you will have some idea of how virtually impossible it is to operate effectively and safely when your arms are restricted.

Imagine, then, how difficult and potentially injurious it is for a horse to be made to jump if his or her rider is using the modern jumping seat, in which the rider often restricts the horse's head and neck by propping his or her hands, holding the reins, of course, on the crest of the horse's neck. This significantly restricts the horse's essential stretching-out of his head and neck over the fence which enables him to make his full effort to clear the obstacle with as little stress as possible on his body.

Riders using this modern seat also invariably throw their weight too far up the horse's neck, out of balance with him, their seats too high and their lower legs often up and back. Despite supporting themselves on their horse's crest as described, this seat is much less secure than the older, classically based seat shown and discussed in Chapter 10. In addition, forcing the horse to haul himself over the fence by lurching over it instead of being able to use his natural, balanced propulsion creates considerable discomfort and potential for injury, particularly to the hind legs, hindquarters and back. The landing will not be as smooth and flowing as if the horse had jumped naturally so the forelegs and shoulders are also at risk.

An eminent UK veterinary surgeon expressed the opinion, some years ago, that from a welfare point of view, no horse should be asked to jump more than 1.5m high—or it may have been less than this—as anything higher greatly risks serious

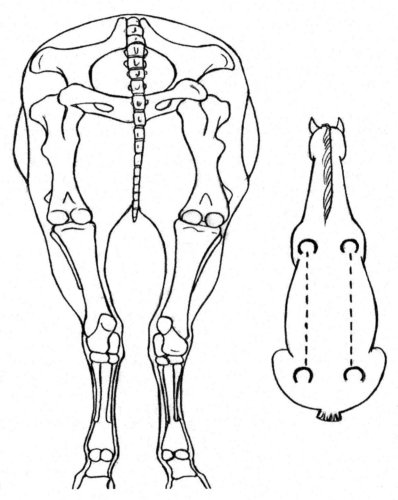

Figure 2.2 The upper leg bones of the horse slant very slightly inwards which brings
the hocks on to the same longitudinal plane as the knees. If this were
not the case, of course, the hind feet would land outside the tracks of the
fore feet instead of following the same line (see diagram). Most horses
are very slightly cow-hocked, too, which means the points of the hocks
point inwards and the hooves fractionally outwards, slightly exaggerated
in our drawing. Some teachers advise students to ride on the outside track
of the manège in slight shoulder-in, to bring the forefeet in front of the
hind, but clearly this is not necessary.

injury to the forelegs. Jumping and cornering at speed, too, clearly is potentially more dangerous for the horse's physique and soundness.

STRENGTH

Speaking generally, the average riding horse in healthy condition weighs about half a ton/tonne. He is said to be able to pull straight backwards, when tied up, with a force of two and a half times his own weight, can kick with his hind legs to a force of 2,000lbs/907kg and the average draught horse can pull a load in a wheeled vehicle (load and vehicle combined) of up to 8,000lbs/3,629kg.

That is some animal. It also stresses how dangerous it can be to be around horses and that effective, humane and stress-free training in general handling and then in riding or draught, as required, is essential for safety, because it usually prevents dangerous, self-defensive behaviour. This is particularly so when you realise, as most experienced horse people do, how lightning fast are horses' reactions to anything that frightens them, and that when panicked they react instinctively, without thinking, ideally by galloping away. My father, as a boy, witnessed a cart-horse galloping in fear, pulling the cart behind him, running straight into a brick wall and killing himself: it was something he never forgot.

Horses' backs, of course, were not designed to take weight from above the spine, only from below it. The heavy and voluminous abdominal contents, also the lungs and heart in the chest, are slung from above, and secured down the insides of the ribs, by various soft tissues. The ribcage, of course, protects the chest contents but the abdominal contents have no such protection.

How much weight in the form of a rider, or a load, should a horse be expected to be able to carry on his back? Some studies have been done on this and the jury is still out, as they say. Generally, the traditional upper limit was that a healthy, strong, sound horse in mid-life can be expected to carry no more than 20 per cent of his own weight, including saddle and accoutrements, less for young, older and unfit or weak horses. Ponies are traditionally said to be able to carry more, but that does not mean that they should be asked to do so, even though, historically, it has been done for centuries by, for example, hill and moorland farmers here in the UK, who would traditionally ride our British native breeds much of the day, for overseeing their land and herding their animals.

It is highly likely that so-called evasions and resistances under saddle are due to the animal being asked to bear too much weight for too long. A common sign of significant overloading is a horse sinking his back and creeping backwards from halt, to relieve his discomfort and pain. Other signs can be reluctance to 'go forward' although the modern hard, sustained bit contact is also responsible for this; squirming; difficulty cornering or working on a curve; halting with the hindlegs splayed or out behind; 'sluggish' gaits and refusing or running out at jumps can result.

SOCIAL NEEDS AND FAMILIES

Horses, of course, are naturally herd-living animals: if you are in a group, there is less chance of your being singled out and killed by a predator, or pack of predators, whose hunger can be sated by the capture and killing of a single animal. This fact of equine life throws up several qualities well-known in modern domestic horses, too, in connection with both riding, training, and care and management.

Most feral horse herds comprise mares and their pre-pubertal offspring of both sexes, with a stallion whose main purpose is to impregnate the mares. Herds may number about a dozen individuals. When colts reach puberty, the stallion usually expels them from the herd. Reports vary on whether stallions will mate with their own daughters remaining in the family with their dams, but there will come a point when a rival, usually a younger stallion expelled from a neighbouring herd and probably living in a 'bachelor band' of young males from about two years old upwards, will try a takeover bid. If this succeeds, it brings new genetic mixes to the family and the old stallion will be expelled, maybe taking up with other ousted stallions or sometimes joining bands of young hopefuls.

Fillies form strong bonds with their dams and relatives and tend to stay in the herd unless horse-napped by a rogue stallion to start or increase the size of his own herd. Horses need company: another need hard-wired into them. Most of them experience stress to various levels if forced to be alone, particularly for long periods. Just because a horse is 'used to' something doesn't mean it is appropriate. Purposely keeping them separate from other horses to prevent them from forming strong bonds which will affect their work with humans is not necessary if they are trained correctly according to ES principles and, for most horses and ponies, would constitute cruelty because of their inherent strong need for favoured company.

Horses gain great comfort and confidence from learning about life with other horses. Their instinct is to follow other horses and this can be a great training boon to teach them to move over tricky ground or off-putting places and over jumps and other obstacles. Friends and relatives ('preferred associates') can be seen grazing with shoulders touching, mutual grooming mainly around the wither area and along the spine. Forcing most horses to be alone, especially for long periods or permanently, greatly distresses them and can trigger ill-health mentally and physically.

WEANING

In feral herds and groups allowed to live fairly naturally in domestic situations, weaning takes place naturally some weeks or a few months before dams are due to foal again, usually in the spring when the grass starts growing to provide rich nourishment for lactating mares, and throughout the summer to nourish growing foals learning to graze. The 'autumn flush' helps animals maintain condition for some weeks or a few months to help get them through the winter, when they will naturally lose some weight. Climate change is affecting the growth of many things, including pastures, so traditional mores regarding nutrient levels cannot be relied on: expert advice is always worthwhile and apps are available which predict nutrient levels in grass on a particular day and time.

In domestic circumstances, weaning is probably one of the most harmful and unkind things humans do to horses, in effect putting foals through a bereavement equal to a child losing its mother long before time. Many behavioural scientists believe that this causes lifelong psychological scars in the foals, which can cause great distress, mental and physical ill-health and consequent problems with training and work when the time comes. Many breeders play this down, if they know it at all, because the conventional method of early and often sudden weaning is convenient for them. Others actually do say they feel guilty, but they 'have to do it'. In reality, early weaning should only be necessary if the foal is exceptionally rough with its dam or the dam is a bad mother. Shortage of milk is easily overcome with appropriate milk substitutes and foal feeds, so the foal can stay with its dam in that case.

As in all aspects of dealing with horses, and other animals, just because something is done that doesn't mean it should be, and just because something isn't done doesn't mean it shouldn't or can't be.

In truth, the whole weaning system and, where breed-appropriate, the dates of foal and youngstock sales, should be hauled up to date so that weaning can take place much more naturally and gradually, and as the horses will dictate. From about nine months of age, the dam is often glad to stop her foal suckling, especially if she is due to foal again in late winter or early spring. Gradual weaning has already begun and can be continued by human attendants. This is a matter of horse welfare and well-being which I feel should no longer be pushed aside and ignored for human gain and convenience.

EYESIGHT

Horses' eyesight seems to be better, but different, than ours. It developed to give the horse almost an all-round view of the horizon: with his head down grazing, which it is for about two-thirds to three-quarters of his time depending on the nature of the food, he can see via a narrow strip of sharp vision, the 'visual streak', almost all around him. He cannot see directly behind him when his head is positioned straight ahead, but with just a slight turn of his head he can see there, too, because, as a prey animal, his eyes are sited at the sides of his head, enabling him to see behind him with one eye. This means, though, that he cannot see directly in front of him close up, where there is a blind spot. Therefore, he loses sight of a jump apparently just before take-off!

The vision outside the visual streak area has about the quality of ours when we see something out of the corner of our eye—serviceable but not sharp. This can make him prone to shying and spooking when he sees something moving that he cannot identify, another excellent reason to train him thoroughly to obey aids/signals/cues for stopping.

Horses graze with a sweeping motion, moving their heads from side to side repeatedly as they walk slowly forward, so they do scan behind them very regularly during feeding. Their long necks which enable them to reach the ground have an added bonus. Horses retain their forest browsing habits but can now reach much higher into trees and shrubs than their smaller, shorter-necked ancestors, so tree leaves as well as grasses are accessible and welcome to them, if available.

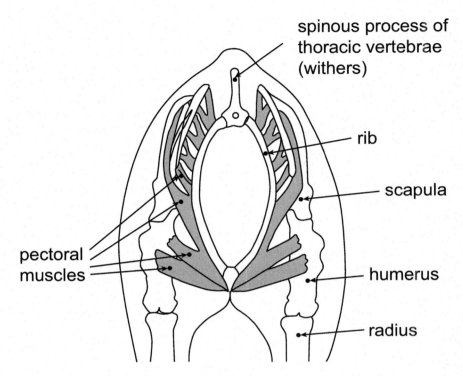

Figure 2.3 View from the front, of the chest area of the horse's skeleton. The horse has
      no collarbone, his forelegs being attached to the ribcage/thorax by means
      of soft tissue structures, which greatly reduces concussion on the forelegs
      when galloping and jumping. This structure is known as 'the thoracic sling'.
      It also gives the forelegs a greater reach and, so, a longer stride, increasing
      the horse's speed. These are ideal survival features for a prey animal.

THE FLIGHT-OR-FIGHT RESPONSE

When feral horses do spot a predator or pack of them, including humans, their reactions are like lightning—something we are all familiar with. Anything that startles
a horse produces the stress hormones cortisol and adrenaline and, in a fraction of a
second, his head is up and he will run, if he possibly can—unless has been gradually trained to suppress this natural instinct to some extent. This is the flight-or-fight
response: horses' first choice in the face of danger is to flee, gallop away, and they can
reach their top speed in a very few seconds. If they can't run away, they will fight, and
horses are *very* strong and heavy (an average riding horse will weigh around half a
ton/tonne) as many of us will know to either our cost or advantage. This innate quality
also makes them potentially very dangerous to us, on the ground or on their backs.

      Once a horse has been badly scared, he can remain alert and hyped up for the rest
of the day because the hormones take some hours to return to normal levels.

      Punishing a horse in any way in such a situation effectively compounds his fear,
which would mean more trouble for ourselves and the horse in the future.

Figure 2.4 The Thoroughbred is often taken as a perfect example of riding horse conformation. This drawing shows a horse standing in the classic photographic pose which shows him to best advantage, and appropriate for any breed or type of horse or pony. The head is turned very slightly towards the camera, regarded as more attractive than looking straight ahead: on the other hand, if the head were turned too much towards the camera it would appear to shorten it and not show the horse in the best light. The nearest foreleg is perpendicular to the ground, with the furthest one slightly back from it. However, the furthest hind leg is as near perpendicular as we can get with a hind leg, and the nearest one slightly back from it. This positioning gives the horse's body an appearance of 'standing over plenty of ground' while also being balanced and showing all four legs. A good riding horse has a slight 'girth groove', as shown here, where the girth will go and, therefore, probably stay, which, with the 'well laid-back' shoulders, good withers (neither too low nor too high) and short back, means he will be able to 'carry a saddle' without the problems of its sliding forward or backward.

HEARING

Horses' range of hearing is similar to ours; they can hear at a higher frequency than we can but do not hear low sounds as well as we have thought they do. The range is between about 50Hz and 30kHz. Voice aids are not allowed in competitive dressage.

Dressage tests were originally used to assess and grade the competence of military horses. If the horses were to be used on a night-time exercise, or any situation in which quietness was essential, the voice would be a give-away as to their position, so the horses had to be trained to respond to signals/aids by means of touch, although they are also adept at responding to our, and other horses', body language.

The original need for voiceless training has long gone, so the previously mentioned dressage rule needs to be rescinded. Horses appear to like the sound of a human voice, depending on how it is used, and certainly respond extremely well to it, being able to learn an amazing number of vocal and other sounds. Dr Marthe Kiley-Worthington wrote that around 200 sounds were quite usual. It is a big advantage to make voice training a clear part of a horse's other work but, like the other aids, or cues or signals as they are known in ES, its use needs to be timed correctly, and used consistently and with the identical tone for each individual cue, for best practice.

Clicker training is popular with many people, but I have never been attracted to it. Trainers in favour of it say that it provides an identical sound, meaning a reward is coming, which horses immediately recognise and so know to expect a reward very quickly after they have given a particular, correct response to a signal. This is true; it's just that a clicker is not necessary for this approach. Clickers can be lost and they take up the use of one of your hands. A single click with your tongue is just as effective, I find (a clicker, of course, makes two), and two clicks seem almost naturally to encourage a horse on to extra effort.

Probably the first thing to learn about using your voice is not to chatter to your horse if you want him to do something specific. We all chat away if 'bonding' with our horse or telling him how gorgeous he is and how much we love him, but training, specific handling and any situation when we want him to do something particular *now* is different. The way to give identical vocal aids is in the identical way for each aid, request or reward each time it is given, and use as few sounds as possible, especially in the early days, to make life easy for your horse.

Horses are brilliant at picking out sounds nevertheless, and you have to be careful not to use requests/commands out of place. A friend of mine, Jane, had a very bright Arab gelding, Jameel, who was always given a piece of carrot when she dismounted at the end of a ride, as well as habitually putting them in his feeds. She always said something like 'here's your carrot' when she gave it to him, not just 'carrot'. One day, she and a friend were standing outside his box chatting when Jane said to her friend: 'I've got to go for some carrots now, I've run out'. Jameel picked out the word 'carrots' in that group of words, left the door where he had been sharing this social event, and went and stood by his manger, one eye on Jane and one ear cocked back towards her. She had no carrots, so she gave him a mint instead. What he thought of that is not recorded. Did 'carrots' to him mean 'carrots' or just 'something tasty?' We'll never know as Jane and Jameel are now both cantering the lands of Trapalanda.

FEEDING TASTES AND NEEDS

The sense of taste is very important to horses. They evolved to eat fibrous vegetation, mainly leaves and sometimes bark from trees and shrubs and, later in their evolution, grasses, plus any other plants which appealed to them, including herbs, the therapeutic qualities of which they may well have learnt to use, maybe through some instinctive drive.

Feral ponies living by the sea eat a substantial amount of seaweed which has several therapeutic properties including thyroxine (for thyroid function) and anti-oxidants. Animals today with laminitis are known, anecdotally at least, to seek out hawthorn leaves. Other resources have been noted, such as wild onion and garlic for skin and respiratory problems, and willow for anti-infective and analgesic purposes. On land that provides an averagely nutritious diet, horses will graze and browse, plus dig for roots, for about two-thirds of their waking hours.

Grassland properties: Many equestrian properties have been developed on land initially used for cattle, without changing the sugar-rich grassland composition to be more suitable for horses. Horses require a more fibrous herbage than cattle, with less sugar and protein than grass specifically for dairy cattle. Horses do better on rougher grazing (apart possibly from Thoroughbred breeding stock) as their digestive systems can break down and use fibre more efficiently than the more complex systems of cattle.

One thing most of us know about horses is that they have a sweet tooth: they mostly love sweet, young grass and leaves and concentrated feeds containing molasses and honey. Sugar lumps, however, are as bad for their teeth, as a treat or reward, as they are for ours, so horse 'nuts' or strips of carrot are better for that purpose.

A taste not easy for horses to find while grazing and one most seem to love is mint. Sugar-free, hard mints are ideal as a reward for desired behaviour as horses respond better to a favourite taste, especially one they do not get on other occasions. It has positive training results provided, as ever, the treat is given *immediately* after the desired behaviour.

No fasting for horses: One of the most damaging and stressful aspects of keeping domestic horses stabled for long periods of the 24-hour day, or even all day, is the usual lack of fibrous feed (hay or haylage). To keep them digestively healthy, comfortable and satisfied, and mentally occupied, they need a more or less constant supply of this bulky, roughage, fibrous fodder. After about two hours and certainly four without it, the vital digestive micro-organisms in the horse's voluminous hind gut begin to die off, possibly triggering mental and physical discomfort and poor health.

Note: Overnight is a particularly bad period for stabled horses because most of them, on observation, have finished their hay by about 10pm, leaving them very much longer than four hours before being hayed up again in the morning. This is a major management black mark against us. If they are bedded on some kind of straw and become desperate, at least they can eat some of that, but an ample hay or haylage supply is the best answer.

Horses are naturally most active around dawn and dusk. They do not sleep straight through the night, as we should, so we need to leave them enough hay or haylage overnight to leave a little still available in the morning. For good doers, it is advisable to take advice from a nutritionist on the calorific content of the fibre/bulk portion of a horse's diet. A lower calorie product may well be better so that adequate fibre and bulk can still be provided.

Because horses are hard-wired to eat for most of their time, it is correct management to ensure that this is possible. Care must be taken to manage them appropriately by providing adequate good-quality fibrous feed to keep them healthy and complement their concentrate feeds, if any. There are high-energy, fibrous, bagged feeds which can be fed alone or mixed with concentrates or balancers to slow down chewing and help to boost the nutrient content of the main hay or haylage ration, if necessary.

Feed before work! Contrary to old beliefs, horses due to work fast and hard must not do so on an empty stomach but need to have a small, fibrous feed up to half an hour before starting work, to help prevent the very acid digestive juices in the stomach splashing up on to the upper part of the stomach and favouring the development of gastric ulcers.

SMELL

It is now believed that horses have as efficient a sense of smell or olfaction as dogs and certainly one much better than ours. Studies have shown that they use their sense of smell to not only identify food, people and other animals but also areas and places at home or away, hormones given off by other animals and people, other horses' rugs, for instance, new objects in their environment and, in stallions, the point of the oestrus cycle in a mare to detect whether she is ready to mate.

Stallions are usually much better at this than their human handlers! Teaser stallions are often used to detect this time, but the serving stallion gets the job of getting mares in foals—which, of course, raises the issue of unkind treatment of the teaser. I remember at one stud I visited they gave the teaser a few mares to keep him happy—and were astounded when his offspring won more races than the main stallion. They promoted him so they had two serving stallions—but brought in another poor teaser.

One of my friends acquired a new horse who had lived with his previous owner for 15 years. He moved right out of the area and was clearly upset but attached to my friend within two days. One night, she forgot her coat and left it hanging just outside his stable door. When she went back in the morning it was on the bedding in his box, flattened and clearly slept on. She found that if she did not leave him something smelling of her to have in his box he would not lie down and rest, and he would always put it on the ground before lying on it. We wondered if he did not like his bedding, and changed it, but the coat behaviour remained the same.

It is a good idea to keep ourselves smelling consistently the same to our horses, keeping to the same toiletries and even the same detergent for washing our clothes, if possible. My first horse was kept on full livery so he was used to different people attending to him. The yard owner one evening told me that, that morning, she had used a different product on her hair and my horse would not let her out of his box until he had thoroughly smelled her head all over and familiarised himself with her different scent. When she changed back to her previous product, though, he did not mind and accepted her just the same.

Horses are notorious for using their sense of smell as a warning of something suspect or dangerous. They can smell a vet several yards/metres away, although the sound of their car engine and their voice are also used for recognition. The previously mentioned horse was due to be vaccinated and the vet arrived in a new car. I was at the other end of the yard and he got to the box before me, but my horse detected him long before he reached the box, even though he had not spoken, and retreated to a back corner of his box, tossing his head.

It is well-known that scents linger, sometimes for days, on objects and on the ground, and horses can follow these trails, especially if left by a friend—or enemy. Scents also seem to trigger memories in horses as they certainly do in humans. I had a horse at livery once and, unknown to us, an old associate of his arrived to live on the same yard. The horse's new owner and I couldn't understand why the two were whinnying and neighing to each other before the new horse actually turned the corner into the yard, but they had clearly scented each other and were delighted to be reunited.

Touch

As highly social animals, the sense of touch is crucial to contentment in horses, yet so much of our management involves keeping horses apart in case they injure each other. The worst form of stabling, for example, is three solid walls and no facility to touch and smell neighbours. This is almost like keeping a horse in solitary confinement which is just as bad, if not worse, for horses than it is for humans, in which species it is regarded as a form of punishment and torture. Horses only being able to see others on the yard is inadequate.

It is understandable that we don't want horses disagreeing and possibly injuring each other over low partitions, or trying to get over the partitions to reach others and risk injury to one or both that way. It has been traditionally a basic taboo to allow horses to touch each other when stabled, but this is so wrong. They need the reassurance of the touch of friends and, provided they get on with each other, neighbours should have at least a small 'chat hole' or meshed window in their partitioning wall.

Some horses, who are friends when out working or in the field, strangely do not get on when stabled next to each other, but this is not common. Conversely, it is not good management to separate friends who *do* get on as neighbours: stabling them next to each other with the facility to both see and touch each other is good for them. If owners find that this creates problems in work when the horses are separated, more attention needs to be given to their training to respond to handlers' and riders' aids, or cues and signals as they are described in Equitation Science, and to their human/horse relationships.

When I was a child, I used to visit an old house with a lovely old stable yard full of intriguing horses and ponies. The stable block was indoor and loose boxes had been made inside by removing stall partitions. Inside, the upper halves of the partitions were vertically barred above strong, wooden lower walls from front to back and, in others, barred to halfway along to the back wall, with the remainder, next to the feeding points, solid wood. This provided two types of stabling according to the horses' preferences.

The fronts of the boxes had been filled in (because stalls were open at the fronts, the horses being tethered at the back where the food was) with strong wooden planking and the usual lower half of a loose-box door. Some of the new boxes had no barring on the fronts, some had a short length of bars and a few had solid wood a short way along the fronts at one end. These arrangements allowed for different horses' preferences and relationships; most could sniff, touch and nuzzle each other and the only two horses who didn't get on were stabled at opposite ends of the block. Ideal.

It is almost par for the course today to fit weaving grilles on stable doors, have solid interior walls and bars filling in the rest of the fronts, over solid lower walls. I cannot think of anything more likely than this to cause horses frustration, loneliness and claustrophobia, other than a box with only an open top door allowing the inmate to look out from his cell. Inside, even people can feel imprisoned, so what must horses feel as creatures naturally evolved to live in the open with no restrictions yet having to endure such closed-in surroundings? Horses generally should be allowed to touch their friends and communicate with them. Obviously, those who don't get on need separating.

A Room with a View

Rear openings in boxes for horses to put their heads out are also a boon and horses love them. They should have a door (ideally made of some see-through, toughened material) which can be closed in bad weather, depending on wind direction.

I am always very sad to see horses turned out on individual, electric-taped postage stamps. Although they can communicate to some extent over the tape if they are very careful, this is no way to keep a horse and is only a little better than keeping them stabled all the time. They are almost as restricted, and the grazing on these patches can be almost non-existent.

Room to Move

I was brought up turning out horses and ponies of all types and ages, and both sexes, together. (Stallions, too, can live happier lives when kept with their 'wives' and offspring.) There were *no* fights but lots of happy horses and this arrangement is good for all ages; youngsters soon learn manners with horses and humans when surrounded by several older mentors. Of course, horses have to be introduced to each other gradually, initially in hand, then perhaps hacked out or ridden around the premises together. Those few who don't like others or cause trouble in a herd can be separated, but some kind of company should, if at all possible, be provided; otherwise the 'outcast' could become lonely and even more of a problem.

SOME INTERESTING PHYSICAL FEATURES

The horse is a quadruped and is, therefore, in a better position to move and balance himself than we are. His stance is more stable, as is his balance. The history of how the earliest equine ancestors lost most of their five toes, being left with just one, which is the equivalent of our middle finger, is a familiar one, as is their gradual increase in size mentioned earlier, and the changes in their dentition, through the various horse ancestors which became extinct during evolution. *Equus caballus* is the Latin name for the horse we know so well, but there are other equine relatives with sub-species—the zebras and asses/donkeys.

Figure 2.5 The horse's head is typical of a prey animal. The ears are very mobile, being able to turn independently and from front to (almost) backward positioning, so picking up sound from all around. The eyes are large and on the side of the head, giving the horse almost all-round vision. If his head is facing straight forward, he cannot see immediately behind him, but can cover that area, too, with one eye, with a small turn of the head. The nostrils are large and mobile, able to expand when breathing at capacity and to inhale scents for identification. The horse's whiskers around the muzzle and the eyes are, in fact, sensory organs acting like antennae and should never be removed; this is now illegal in some countries.

Donkeys may have been domesticated before horses and still experience appalling working lives in some countries. Zebras are widely regarded as 'untameable' yet were once not uncommonly seen on circuses in liberty acts and have been trained to harness (one Victorian gentleman having used his zebras and carriage to transport himself around London, although what his coachman thought of them is not known). They have been trained to saddle for riding, including in the English hunting field, side-saddle, including jumping, by a certain, intrepid Victorian lady. The tall, elegant Grévy's zebras with narrow stripes do not get on with each other and are territorial but Plains zebras are wanderers, like equids, and want each other's company. The rarest zebra is the Cape zebra, with the distinctive gridiron pattern on the hindquarters. The South Africans, understandably, won't let any of them out of the country.

BENEFITS OF THE EQUINE SKELETON

All animals having an internal bony skeleton share a common skeletal foundation, with certain species differences. Horses' skeletons have no clavicles/collarbones so their shoulder blades and forelegs are linked to the front part of the ribcage by large, strong muscles and other soft tissues. This evolutionary adaptation, known as the 'thoracic sling', means the forelegs and shoulder blades are not restricted by a collarbone, thereby enabling a longer stride and, so, faster speed. The muscles and soft tissues absorb more of the impact on the shoulder joint and the joints of the foreleg caused by galloping and jumping than would be possible with a collarbone.

An interesting feature, yet again helpful in a prey animal, is their 'stay apparatus', giving them the ability to 'lock' the joints of their legs when sleeping lightly standing up, so that they do not fall over and can make a very quick getaway should danger threaten. It takes a horse lying down, particularly flat out, a few seconds to wake, get to his feet and into a gallop, which can make the difference between life and death. The locking is done by means of a collection of muscles, tendons and ligaments fixing the joints of the legs. It allows the horse to brace the forelegs and, often, one hind leg, resting the other on the tip of the hoof, and alternating which leg rests.

I was brought up in Blackpool where my parents had a hotel, and the harness horses giving visitors landau rides up and down the Promenade would, and still do, often nap like this between fares, or eat from the nosebags their drivers put on their heads. There are water troughs at certain points along the Promenade, the top level for horses and the lower one for small animals.

Horses sleep in short naps, for around five hours out of 24, although stabled horses, subjected inevitably to a degree of boredom, seem to sleep for around seven hours per 24. Horses sleep either propped up on their breastbones or flat out. When flat out, they sleep deeply for only up to half an hour at a time because their considerable weight interferes with blood circulation. If horses are weak because of malnutrition, arthritis or old age and the like, they often start to have trouble getting up after lying down (in either position) or rolling and might be seen sitting up and swivelling around on their bottoms. A vet explained to me years ago that if they can't get up and have to lie down flat from fatigue again, the lung on their ground side can fill with

Figure 2.6 The horse naturally spends a good two-thirds of his time with his head down grazing his most important food. The eyes are high up on his head above the ground, giving him a good all-round view, and the legs, being relatively thin, barely obstruct that view, so he can keep an eye out for predators. The horse's tail grows full down from the dock and is to protect the horse from adverse weather from behind (when horses will stand with their backs to the wind and rain) and use the tail as a valuable fly whisk during warmer weather to dislodge insects. The fashion for pulling tails obviously deprives them of these important functions. The author's preference is for 'racehorse tails'—full at the top and banged (cut level) across the bottom.

body fluids and not clear because of their weight, and they can slowly drown, so if you spot a horse having problems getting up, call a vet at once.

SOME THOUGHTS ON CONFORMATION

The phrase 'a well laid-back shoulder' is often used to describe good riding-horse conformation, but what does it mean? Generally, the spine along the shoulder blade is best at very close to a 45° angle to give the horse 'a good front' and the rider the feeling of 'plenty in front of him'. Also, for speed, reach and ease of movement, a horse should have an 'open elbow'; that is, you should be able to fit your fist into the space between the elbow and the ribcage. The opposite of this is known as a 'tied-in' elbow.

In the hind legs, as well as 'open' and clearly defined stifles (the equivalent to our knees), a point to look for is what are termed 'well-let down' hocks. This means that the hocks (equivalent to our ankle joints) should not be much higher than the knees (equivalent to our wrists), with short cannon bones below both. Horses are naturally *very* slightly cow-hocked (their hocks turn inwards) but, seen from behind, their cannon bones should be straight in all four legs and in the same plane, for future strength and soundness.

From a riding viewpoint, a horse's withers should not be lower than his croup. If they are, he is known as 'croup high' and will always give his rider the unpleasant and slightly insecure feeling of going downhill. Such horses can be improved as to their balance and way of going with classical training, the results of which will give the forehand a measure of 'lift', although nothing will entirely make up for this conformation fault.

Having said that, many Thoroughbreds of sprinting strains now have noticeably croup-high hind quarters with long, powerful hind legs and strong muscling. This is a trait unintentionally bred into American Thoroughbreds which became obvious during the 1960s and 1970s, the Americans placing a premium on speed above virtually all else. A good example of evolution in action producing a horse for the job. The Thoroughbred now being a truly international breed, having been 'assembled' in Great Britain from mainly foreign 'parts', or breeds, this croup-high feature in sprinting Thoroughbreds with American ancestors is seen worldwide. Such horses, however, sometimes move with a 'rhumba'-like action because they have to place their hind hooves outside the line of their front ones, instead of in their tracks or over-tracking conventionally, to avoid treading on their own fore heels. This movement would not be popular with dressage judges, but such horses are interesting to ride.

CORRECT VERSUS 'NORMAL'

The 'lifted' forehand in a correctly schooled horse going in true collection *on a light bit contact* is a sure sign of a rider/trainer who knows what they are doing. This correct training develops the horse's muscles to best effect, those he uses to move correctly and classically, and which thereby protect the forehand and legs. This does not happen in horses forced into a false forehand posture because the muscles are not developed correctly. To uninitiated onlookers, or those inured to it because it is so common today in what is often known as English-style riding, this posture might look as though the horse is collected, but in practice the head and neck are usually

Figure 2.7 Freedom and fun. Most horses love a good roll and it is good for mind
and body. Not only is it enjoyable but also it helps to clean the hair,
believe it or not looking at this picture. Mud helps to give the skin a
health treatment and, if plastered on thickly enough, protects horses from
the wind and sun. The author visited a top Thoroughbred stud many years
ago and was intrigued to watch millions of pounds (sterling) worth of
broodmares and foals being led in for evening feeds. On commenting
about the huge task of brushing off the mud, the Head Stud Groom said
that they did not brush it off as it protected the horses from the weather;
also the mud was left in the horses' feet till the following morning so that
they would not become packed with droppings overnight.

being pulled in by the rider's hands, shortening the neck, cramping the throat so
interfering with the horse's breathing, and preventing him from swallowing his own
saliva, hence the copious froth produced by some such horses.

These effects of force are added to if the horse is also going with the front of
his face behind the vertical. This significantly limits his forward vision unless he,
uncomfortably, moves his eyes upward in their sockets all the time—'looks up'.
The BTV (behind the vertical) posture and his poll not being the highest point of
his outline, as it should be, but bowed down so that the highest point of the outline
comes about a third of the way down the neck, are so common today that they are
regarded as normal. The effects of these training/riding faults on a horse must be
truly frightening for him.

Sometimes, this habitual posture produces a semi-permanent shape called a 'broken crest', where that point actually has an angular appearance. With at least several
months' rest from the faulty riding that brought about that temporary deformity, the

horse's neck conformation can become normal again provided any damage to the neck vertebrae has not been too severe.

The effect of correct training is that all the appropriate 'riding' muscles in the horse's body are developed, not least those in the hindquarters which need to be strong to bear the extra weight carried there because of the slight rearward shift of the horse's balance in collection. This clearly is also of considerable benefit to jumping horses, enhancing the power and thrust from the stronger hindquarters and hind legs. Classical techniques, just like ES principles, benefit all horses and ponies in all types of work and can compensate for many moderate conformation faults.

Managing a Rider

The horse's back comes in for a good deal of stress and strain during riding, so it really is important that riding horses are not hampered and potentially injured by poorly fitting saddles and over-heavy riders, and that students are taught the necessary skills of balance and control of their bodies, which are dealt with in Chapters 7 and 9. Sitting to the trot and canter can be hard for novices to master, so I hope the tips given in those chapters will help with this problem.

It causes considerable discomfort and distress to horses when a rider is banging up and down on their backs and swinging from side to side, and it is essential for riders to master their own balance and control if progress is to be made, not least because riders otherwise often hang on to the bit, via the reins, to stabilise themselves and cause their horses significant pain in their mouths, and confusion, as well. Both these situations represent welfare issues, and confusion, in particular, due to random pressures in the mouth, is a prime cause of defensive 'misbehaviour' in horses.

We train horses to go in the classical way because it is the best way to develop their bodies so that they can use themselves optimally and naturally and carry a rider in all gaits and over obstacles with the least effort and risk of injury because they have been correctly strengthened. This makes for a stronger, calmer, happier and more trusting horse, with the enhanced physique, action and performance of the correctly trained and impressive riding horse. The rider is more secure, and therefore safer, riding in the classical seat, with appropriate adaptations for speed and jumping. He or she needs to expend less effort, and finds partnering such a horse, particularly if he is home-trained, rewarding and often exhilarating.

Equitation Science makes it even more feasible to get through to our horses in a humane, effective and, to them, much more comprehensible way than any method yet devised simply because, as stated earlier, ES is not in itself a method or 'school' of thought but a set of proven principles, universally applicable, presented in a rational order relating to how horses think and learn, making life better for horses and humans alike. It continues to develop and reveal new information and advice for the benefit of both species.

There is always a risk in any modality of people putting their own spin on it and, even in true classicism, we do find some, if minor, differences of opinion which usually come about because of horses' individuality. In ES, too, very slight differences may occur, I find, for the same reason, but here we do have a rigorously researched and planned rational training programme and a 'route map' along which we can retrace our steps to an earlier stage, if we have problems, and so find a solution to them.

EQUINE LEARNING THEORY

When managing, caring for, handling and training horses, it is clearly important that all the previously mentioned factors presented in this chapter need to be taken into account, bearing in mind that we also need to keep up to date with research and developments in the field that might change things.

> Learning theory is an explanation of how students, in our case horses, take in, process and use in practice what they learn.

So far as horses are concerned, they have been seriously hampered probably since their domestication by humans naturally having assumed that they think and learn like humans—indeed, we have applied this *modus operandi* to virtually all other species. Now, though, so much research has been done on the equine brain, behaviour, motivations and survival traits that we know that that approach is inappropriate and has probably caused horses more regular and lasting grief than anything else we have done to them. Humans are in general reluctant to accept change even when a new version of an old idea is clearly better. I hope that reasons for change and moving on in a better way will become apparent in this book and will result in improved well-being for our horses and more satisfaction and enjoyment for us.

IN A NUTSHELL

Of all the animals humans have domesticated, the horse probably has been the most useful and important. Human civilisations almost certainly would not have developed in the way they have without horses to support their growth.

The horse's evolution has produced the characteristics, qualities, propensities and needs of our modern horses and we have considered and assessed these in regard to their relationship with humans. We have looked at basic training principles, interesting physical features, conformation pointers, the horses' problems in carrying a rider and how things can be improved for them.

Equine learning theory, on which Equitation Science is based, has been described as an explanation of how horses take in, process and use in practice what they learn, intentionally and unintentionally from the human viewpoint. One of the worst mistakes made in horses' training and management is for people to believe that horses think as humans do and treat them accordingly.

3 Recognising Equine Emotions

At the end of Chapter 2, I mentioned that we humans have, down the ages, treated horses and other animals as though they think and learn like us, but now it is becoming increasingly obvious that they do not. The same can be said for the emotions that horses have and express. We are often so ready to make human judgements on equine behaviour without really thinking through what is before us and why it might be happening. Sometimes, let's admit it, we put a 'meaning' on a behaviour or expression that reflects what we should like it to mean rather than what it does mean.

Horses are very sensitive, perceptive animals with senses not quite like ours (already described here and in my book *Horse Senses*). We have no reason to think that they feel pain less than we do, but it is known that they are very good at disguising it because, as prey animals, the sick and injured are usually compromised and not so able to get away to safety and follow the herd when predators threaten. Predators, for their part, depend on making fairly regular kills to survive and bring up their young; so they have become excellent at spotting prey having problems, moving strangely, having trouble keeping up or even just being very young or old, because they will be easier to catch.

Even those who have equine welfare at heart have made mistakes in translating equine body language and interpreting it, often missing signs and postures which are increasingly being explained and revealed for what they really are rather than what popular conception in the horse world has believed up to now.

Here in the UK, in conjunction with Dr Sue Dyson, the Saddle Research Trust has been doing excellent work into the expression by equines of their emotions, and there have been some surprises.

GENERAL APPEARANCE OF GOOD HEALTH, CONTENTMENT AND RELAXATION

SIGNS OF GOOD HEALTH

A healthy horse with a positive outlook on life, content and feeling good will have certain well-known features about his appearance although, because horses live very much 'in the moment', this can change quite quickly.

A healthy, content horse's first immediately obvious characteristic will be a naturally glossy, soft and smooth coat, perhaps springy to the touch if he is not clipped, and with a mane and tail with similar characteristics. This applies even if the horse is roughed off at grass. In fact, 'Dr Green' has traditionally been known to be a real pick-me-up and health-and-happiness booster, other conditions being good, such as clean water, company and shelter from insects and extremes of weather.

DOI: 10.1201/9781003121190-3

Figure 3.1 This Arabian stallion's head exhibits the valuable survival features of the equine head—mobile ears, large, side-placed eyes and mobile nostrils. This head attitude displays alertness and interest in something some distance away, the head being lifted to give the eyes their best focus on distant objects, the ears pricked forward to pick up any helpful sounds and the nostrils flaring to detect scents.

Most people would then look at the horse's head and face and note the general demeanour which will say clearly how he is feeling. Bright eyes, neither sunken nor wide open in alarm, are always a positive sign of good health as is an interest being paid in visitors and the surroundings.

Figure 3.2 This photo shows a defensive threat attitude. The horse looks more afraid than angry but is warning off the human. The damage he, presumably, has done to his stable door in frustration shows that his management is lacking in one or more basic essentials.

The horse's general body stance and movements will look relaxed and comfortable, and effortless whether getting up and lying down, galloping around with friends, wandering and loafing or grazing with head down and walking all the while. If there are high browsing facilities such as trees and shrubs, the horse will be able to reach up to his choice of leaves without difficulty and be able to chew and drink easily. Horses lie down more and eat a little less when stabled but these basic, general signs of good health and feeling good are the same.

CLOSE OBSERVANCE

Because horses are so good at disguising their negative states, any suspicion of difficulty in moving needs investigating, such as lameness or slight unevenness, trouble getting down and, particularly, up again, not lying down to sleep despite the stable's being large enough and the bedding comfortable and clean, half-hearted rolling, and also problems masticating food of all types, difficulty in swallowing food or water, reluctance to eat or drink and leaving the normal rations, although the quality of the food and/or water could be the problem here. In short, *any* abnormal behaviour or appearance for the individual needs investigating. Staying apart from the herd rather than joining in is usually a clear sign that something has changed, maybe in the herd dynamics, or that the horse concerned is 'not right'.

The appearance of a horse's droppings and urine depend largely on his diet but can be affected by illness or malcontent and/or a stressful lifestyle. A horse will pass between eight and 12 piles of droppings a day of a green to khaki colour. The individual balls break easily on hitting the ground. If your horse's droppings are looser (some level of diarrhoea) or smaller and harder than usual (possible constipation), there is a problem. His urine can be clear or cloudy and of a pale yellow colour: watch him, or her, to see if the process of urination seems easy or difficult, painful or erratic.

MORE SPECIFIC POINTERS TO GOOD HEALTH AND EMOTIONS, OR OTHERWISE

A good deal of understanding horses is reliant on the sensitivity and perception of the person involved, not to mention their attitude to the status of horses in our society: some have much more empathy and innate 'feel' than others, some care about their horses' well-being but some don't and, as mentioned, it is simple to underestimate or wrongly identify signs or just pass off a hunch that all is not as it should be. If you feel that a horse is just 'not right', he may well not be. The more you can open your mind and your heart to your horse, study horses in general and in detail, and really try to look at life the way they do, with our up-to-date knowledge, the more likely you are, over time, to develop 'horse sense'.

WHAT DOES IT ALL MEAN?

Often, an experienced horse person, particularly with a horse of his or her own or another with whom they are familiar, can tell at a glance that something is amiss with a horse. Often a horse out of sorts will look withdrawn, the eyes will look sunken and dull, the ears loose and held rather sideways, the head held fairly low. The skin might look and feel 'tight' and not move easily under the flat of the hand, and the coat may look dull and be 'staring' (standing up) or stiff. The horse will probably show little interest in his surroundings.

If a leg is being held either in front of or behind its vertical pair, it could be painful, the tail could be hanging between the buttocks instead of being held a little away from them and the horse could have an overall depressed or unwell look about him. Obviously, if the horse is shivering or sweating, either a warm sweat or a cold one, there's a problem. A good, up-to-date veterinary book on health and first aid for horse owners is an invaluable must—ask your vet to suggest one—and the temptation to follow free advice from social media can be too dangerous to risk, in my view.

EMOTIONAL SIGNS

Basic signs of equine emotions were given in Chapter 1. Although horses are good at disguising when they feel out of sorts, they do show their emotions very plainly to those who understand the signs, particularly other horses! Some people, though, do miss those signs by not noticing them or misinterpreting them. Sometimes an interpretation centuries old turns out to be wrong with our improved knowledge of equine behaviour.

Figure 3.3 Friends, freedom and foraging. These two playmates have access to all three of these equine essentials. Although we shouldn't want to be in the middle of this interaction, horse play can be rough and they clearly enjoy it.

For instance, it is common for a horse whose ears are flattened back to be said to be concentrating fully on his rider, whereas we now know that it is a sure sign of anger, significant distress and/or pain. Showing the whites of the eyes is generally unreliable as a judgement of emotion and may well *not* mean that the horse has an unpleasant nature; it may simply be the structure of the eyes if they show a good deal. If they show rarely, however, in a particular horse, when they do show, it can indicate fear. The triangular eye posture or 'pointed' shape of the upper eyelid indicates anxiety verging on fear and, as previously mentioned, a dull and sunken appearance indicates ill health but can also point to unhappiness, depression and being generally mentally out of sorts.

A combinations of signs can be helpful and you can get a lot of practice at this at any equestrian event by studying the horses not only while competing but also warming up, cooling down and generally on the show or competition ground.

If you see a horse with his ears pricked hard forward, wide eyes, flared nostrils rushing his fences with, if he can manage it, his head high and maybe his tail held high, he is not keen to get at the jump but frightened. A horse who has napped/baulked before a fence probably has ears back, an angry look in his eyes, flared nostrils and a swishing or thrashing tail. With today's fashion for tight bridles pressurising the

Figure 3.4 The calm, interested expression on Zareeba's face shows that he loves his pussy cat friend. Most horses prefer other equine company, but many make firm friends with animals of other species.

Figure 3.5 This Arabian stallion lived in and out with his wife and their offspring, perfectly happily. Few domesticated horses are given the chance of such a natural, happy lifestyle.

base of the ears, too-high bits pulling up the corners of the horse's mouth creating wrinkles, nosebands so tight they leave a ridge around the horse's jaws or even cut the skin, incorrectly adjusted double bridles and curb chains, the horses have every right to display pain and anger.

Pain is shown by anxious eyes (described earlier), ears hard back, nostrils wrinkled up and back, opened lips, grinding the teeth, champing at the bit actively (rather than just mouthing it which is impossible with a tight noseband), a lolling tongue (which a tight noseband usually prevents depending on its design), excessive froth as mentioned, tossing the head, swishing or thrashing the tail or, often if the pain is in the back or body as opposed to the head, holding the dock stiffly so that only the loose hair below it is moving, and raising the head somewhat. Obviously, if a horse is clearly injured or lame, several signs of pain will be present.

The ears do indicate where the horse's attention is. If they are pricked forward but not held there continuously (which indicates fear), it means interest and curiosity. Ears that flick independently to and fro indicate the horse taking in several stimuli at once and the direction in which the ears are pointing will show where they are. Horses led in hand quite often hold the ear on the side of the leader towards him or her. The nostrils should normally widen and relax softly with the horse's breathing: if they are constantly flared the horse could be breathing hard due to hard work, he could be having breathing difficulties or he could be frightened.

His general body demeanour also indicates his emotions. The familiar 'upside down' posture, with back hollowed down, head and neck held up (tack and rider permitting) with ears pricked and nose poking plus wide eyes and flared nostrils, of course show that the horse is very excited or frightened. The tail will be held up and maybe stiffly out, the hind legs held and used too far back, the gait 'paddling' and quickened ready for flight, and the foreleg action may be higher than usual. The whole body will be stiff and ready to go.

On the other hand, a horse relaxed and going well will be using his hindlegs probably brought forward under his body, his dock will be swinging rhythmically (not just the loose hairs below it), his back will be slightly raised although relaxed enough to make its natural movements in action, and his forehand action clear and natural. If he has a real classical rider or one knowledgeable about ES, his neck will be voluntarily and naturally stretched forward, not positioned by his rider, the rider following and staying in touch with his mouth lightly. The neck posture will depend on the horse's level of training—in an advanced horse, it will be stretched up and forward, or lower and out in a more novice one. The head will be carried probably just in front of the vertical with his poll the highest point of his outline—the stipulated correct way of going which is rarely regarded by contemporary trainers, judges and therefore riders.

A RESULT OF GOOD TRAINING

This free yet voluntarily controlled way of going comes as a result of good training and a horse who gives every impression of going happily along with what he is being

Figure 3.6 It can be difficult to tell the difference between alarm and fear, the former sometimes leading to the latter. Here, we show the attitude of a frightened horse, with his head high, ears pricked towards some 'monster', nostrils wide and probably the whites of his eyes would be showing, too. His body will be stiff and his tail clamped between his buttocks. The head study shows pricked ears, wide eyes and nostrils and stiff muzzle, all signs of tension preparing for flight and fear aimed at gathering as much sensory information as possible regarding the perceived threat.

asked to do—in other words, responding calmly, if with some panache and enjoyment, to the cues/aids he is feeling and knows how to respond to. Frightened, angry or anxious horses never go like this.

HAPPY AT HOME

If a horse has a background life of security and contentment, he will be confident in his life, curious about his surroundings wherever he is, trusting of his people (even the vet!) and show all the signs of calmness, relaxation and sheer well-being. This does not mean that excitement on occasion is a bad thing but that he can confidently express emotions, and his rider can also be confident that the horse is extremely unlikely to show any of the self-defensive behaviours such as bucking, kicking or striking, napping/baulking, shying, rearing, running off or even bolting or 'leaving the arena' without permission because he has no need to do so.

EQUINE MENTAL HEALTH

Confusion, usually from having to tolerate opposing aids/signals given at the same moment, sustained pressure from bit and legs/spurs and incorrect negative

Figure 3.7 The horse with pricked ears, rounded but soft eyes and rounded nostrils shows
alertness and interest, but an alert horse will not be tense and maybe trembling,
as are commonly present in fear. The drawing with ears to the sides, soft eyes
and relaxed half-open with soft muzzle shows a relaxed horse.

reinforcement in general (*see* Chapters 6 and 7), is a major problem in modern, conventional riding. Horses respond in two main ways. They either exhibit the kind of
behaviour just mentioned or they sink into what is termed 'learned helplessness',
which basically means that the horse has learnt that he is helpless to improve his lot,
he can't do right for doing wrong, doesn't understand how to escape the pressures
applied to his body or other stimuli such as vocal commands and just gives up.

Some scientists feel this is a form of clinical depression in which the horse can easily be mistaken for being ill. In fact, he *is* suffering from seriously poor mental health
and, even with a huge and horse-friendly change in his routine and lifestyle, it can be
very hard to bring a horse out of it, although significant improvements can be made.
I find this state is more common in still-working and former competition horses (at
all levels) than racehorses, although some of the latter can be affected. I find they are
more prone to stereotypical behaviours ('vices') than learned helplessness.

In horses who cannot touch and smell others for large parts of their day, and night,
loneliness can soon set in, with all the same adverse effects that it has on humans. Simply being able to see other horses next door or around the yard is not enough, contrary
to the still-prevailing, archaic view that horses should not be able to touch and sniff
each other in stables. Obviously, horses who clearly don't like being next to each other
should be well separated, but physical and mental health requirements stipulate that
those who do get on need to be able to at least touch and sniff each other in stables.

The still common practice of enforced stabling, which can often be a fairly permanent arrangement in yards with inadequate facilities for freedom and exercise or mistaken, outdated opinions of correct, or even merely acceptable, equine management, can result in boredom and poor mental and physical health, as horses desperately try to find something with which to occupy themselves. I am sure many horses suffer from claustrophobia (not surprising in view of their evolution). There is often a lack of hay or haylage in these situations and, combined with not being able to adequately socialise with other horses, this can significantly reduce a horse's well-being. Remember the Three Fs—friends, freedom and foraging—which should be available daily and for many hours out of the 24. These are what horses need most.

THE ETHICS OF MAKING HORSES WORK

There can be no doubt that all horse training and just about everything we do with horses labels us as users, exploiting their very considerable attractions of strength, speed and trainability. Sadly, in many cases, this can be exploitation of the worst kind, with no consideration for the nature of horses, no attempt to allow bonding to take place not only with their own kind but also with us, their handlers, trainers, riders and grooms. Two excuses for continuing poor management policies are that it is the most convenient method for their human 'bosses' and that the horses are used to it. Neither is relevant when considering the well-being of the horses.

At its best, though, the life of a domestic horse can be, to all appearances, enjoyable and less demanding than that of his feral cousin, exposed as feral horses are to all the vagaries of weather, predation in many areas where there are feral herds, no help of any practical kind when injury, illness or foaling difficulties arise, no food or water at times and possibly no protection from the severe attacks of insects. To top that off, in some countries feral horses are hunted and shot to reduce their numbers, harassed by helicopters, herded into human facilities and families split up, although the latter is also the norm in domestication.

AIM AT BETTER

Some of these things can also happen to badly managed domestic horses, of course, but, done properly, domestic horse care and management can provide horses with a safe, comfortable, interesting, fulfilling and rewarding life. I agree that that sort of life for domestic horses is in the minority, but we can aim at it, and the closer we get to achieving it, the better.

Some people do believe that breeding horses for our own pleasure is unethical, horses no longer being essential in at least one half of the world for transport, farm work, business or industry, but there is no doubt that well-kept and used horses can have a good life and enjoyable or at least tolerable work. They can enjoy being with each other, bonding with us and experiencing variety in their lives, including their work, which is welcome to many of them.

The problems come when making money, accumulating kudos and winning prizes take precedence over equine welfare, and those situations are extremely common. I believe that it is perfectly possible to do those things, if someone really has to, and still give the horses a good life. Separations can be as painful for horses as for humans, and they are a seemingly unavoidable part of our modern horse world. Horses have elephantine memories: they remember and greet old friends, relatives and associates sometimes many years after being separated.

I cannot see a way round this source of distress to our horses. It is said that horses live almost *entirely* in the moment, but I am not sure about this. We can certainly train on that basis, and it works, but I find it hard to believe that, if they remember old-time associates, they do not at least occasionally come into their minds. But perhaps we shall never know.

THE FIVE DOMAINS

As for everything else involved in a domestic horse's life, the qualities of a horse-friendly life are well-known and clearly set out in the Five Domains of Animal Welfare (*see* Chapter 4), formerly known as the Five Freedoms. The peoples of most democratic western countries abhor excessive legal regulation, and goodness knows we have plenty of it, but there is room for a good deal more of it in general animal

Figure 3.8 The drawing of the whole horse shows the typical stance of a horse in pain—ears back or to the sides, head lowered, tail raised and maybe swishing or thrashing, body could be stiff and tense. The head study shows the facial expression of pain—ears back, eyes more a triangular shape, nostrils pulled up and back, maybe the mouth open and teeth showing in a grimace. The horse may or may not be groaning and pointing to or snatching at the site of the pain with his muzzle, or kicking at his belly if he has colic. With abdominal pain, he may keep getting down to roll and then getting up again.

management, particularly to help those animals not in the public domain in unregulated premises and private properties.

As for the future of animal use in sports and hobbies, there is no doubt that, with the increased public interest in the animals' well-being, some sectors of the horse world need to stop merely paying lip service to it by claiming that it is paramount in their dealings with their animals and do a lot more to improve it. Those who don't even bother to claim that also need to up their game and implement appropriate practices and standards according to the animals they keep. If these improvements are not made, others will make them for them or put a stop to their activities.

IN A NUTSHELL

We know that horses are more sensitive emotionally and physically than was thought in previous generations. It is becoming clearer that they do not, after all, think and process information and emotions as we do.

In this chapter, detailed information has been given about the physical and emotional signs shown by horses in relation to their health and well-being, and more accurate information has been given as to what they really mean and how to interpret them. Combinations of signs and postures can be very revealing, particularly whole-body postures. The beneficial results of good training and a happy home life are explained and an important section on equine mental health follows. The Three Fs—friends, freedom and foraging—are briefly introduced as being crucial for health and happiness. Finally, the ethics of making horses work for us are considered.

4 Happy at Home

A happy home life is one of life's greatest treasures, and this applies to animals as well as humans. Let's suspend for a while the belief of many scientists that animals do not experience happiness as we know it. Instead, let's think along the lines of trying to create and maintain a quality of life at least approximating to happiness for our horses and any other animals we have, whether we think we use animals or not and whether they are working, retired or play a valuable role as a friend and companion.

It is acknowledged by professionals involved in the health care of humans and animals that sustained contentment or happiness or, conversely, misery, discomfort, fear and a feeling of helplessness can actively affect physical and mental health. 'Happy hormones', particularly oxytocin, are produced and circulate round the body when we, and our horses, are living a life that is appropriate to us and makes us feel good: this promotes and helps sustain good health of both kinds. When conditions are more negative, though, in the 'down' sense, the opposite can occur.

Gastric ulcers are a well-known effect of stress, unhappiness and an inappropriate lifestyle. I first heard of them when my father was diagnosed with them when I was a teenager; despite treatment, he had them for the rest of his working life, after which they just disappeared. We hear that the majority of racehorses and competition horses are also affected by gastric ulcers, plus some others kept as pets, family hacks or companions, if they are living a life that is not suitable for them. Even foals can get them and, unsurprisingly, they have also been diagnosed in weanlings and yearlings whose main problems start with the very distressing, early and artificial weaning to which we subject them. Poor health in general *can* stem just from living conditions inappropriate for the individual.

Behavioural problems, from minor schooling/training failings to really dangerous ones like rearing, bucking, bolting and shying, and stereotypical behaviours, formerly called 'stable vices', can result from horses being kept inappropriately but also from their not having 'agency' over their lives—feelings of control, prediction, free will and a general feeling of having some choice over their daily lives and activities.

HIS INNER SELF

One area of our existence, ours and our animals', may not be entirely suitable for a book from an academic publisher, but I feel it should be mentioned and that is the third of the trilogy usually quoted as 'mind, body and spirit'. This is nothing to do with religion or even with a 'belief system' (I really dislike that phrase) but with our selves—our spirituality.

People who have been around animals most of their lives, if they have any sensitivity at all, must, surely, be aware that they, like us, do have a sense of self, an

DOI: 10.1201/9781003121190-4

awareness of not only their environment and external lives but also a kind of inner knowing or acceptance of, perhaps, a sixth sense. Maybe I am not describing this very well, but I think many of us must be conscious of its existence, though it would be probably impossible to prove to the satisfaction of science as it is not measurable or definable. Anyway, without becoming 'deep' and airy-fairy, I 'know' spirituality exists in both humans and animals of most kinds. It could well be what produces our hunches, tells us and animals whether we are 'in a good place', whether something is right or wrong for us regardless of whether or not we like it. It could also work in tandem with our and our animals' genetic propensities in creating our individuality.

So, this chapter is a holistic one, encompassing the needs of a horse's mind, body and spirit, so being in accordance with the shared ethos of classicism and Equitation Science, which goes far beyond training, handling and riding.

So very many horses are now kept just as competition vehicles and kept in a way that is most convenient for their human connections, regardless of what the horses

Figure 4.1 Ample turn-out time (as many hours a day as can be arranged, and as the horse shows he needs) to let off steam, enjoy freedom and perform normal equine high jinks is vitally important to horses and ponies, ideally with company.

themselves really need. Often, this makes for substandard care so far as health, welfare and well-being are concerned, although their human connections claim that their horses have 'the best of everything' (otherwise they wouldn't be able to do the work required of them!). In practice, such horses rarely have the most basic needs of a horse fulfilled—those Three Fs, tactile contact with friends for moral support, freedom for the benefits of company and almost constant movement, and forage and foraging for their nutritional and occupational value. If they did, how much happier and, surely, healthier they would be and how much better they would do their work.

THE FIVE DOMAINS OF ANIMAL WELFARE

Let us look at an excellent statement of animals' needs, updated from the former Five Freedoms of Animal Welfare which stated that we should aim to keep them free from hunger and thirst; discomfort; pain, injury or disease; allow them freedom to express normal behaviour; and finally, they should have freedom from fear and distress. The Five Freedoms broke new ground in presenting clear baselines for animal welfare and well-being. The Five Domains put a directly positive spin on them by stating what animals actually need in practice for a good life.

Horses are very adaptable but only within individual limits: conditions outside those limits will probably cause happiness and health to suffer. It is still true that horses are individuals and should be recognised, assessed and treated as such within a horse-appropriate framework. Let's look at the Five Domains of Animal Welfare as a statement of intent on our part to create for our horses a happy and nurturing home life, while bearing in mind the original wording of the Five Freedoms.

The domains are categorised as Nutrition, Physical Environment, Health, Behavioural Interactions and Mental State. There are slightly varying definitions of what is recommended in each category and you can explore their principles and ethics online. The following list gives the salient points at the time of writing, pending ongoing considerations:

1 *Nutrition:* Animals should be ensured access to a diet that is sufficient, balanced and varied within the requirement of clean food and water.
2 *Physical environment:* Animals should be made comfortable through environmental temperature; substrate or ground/floor conditions; adequate space; air quality; odour; noise; and predictability.
3 *Health:* Good health should be aimed at through the absence of disease, injury and impairment, and maintenance of a good fitness level.
4 *Behaviour interactions:* Animals should be provided with varied, novel and engaging environmental challenges through sensory inputs, exploration, foraging, bonding, play, facilities to escape harassment, and to sleep and rest adequately, plus other factors.
5 *Mental state:* By creating positive situations in the previous four welfare elements, the mental state of an animal should be enhanced by experiencing mainly positive states such as comfort, pleasure, curiosity and vitality while minimising exposure to detrimental states like hunger and thirst, pain or discomfort, fear and anxiety, loneliness, fatigue and exhaustion, frustration or confusion, and boredom.

From the previous points, it will be clear that fulfilling all those recommendations may not be easy, but they should all be able to be met and perhaps with not too much difficulty, and with a little ingenuity and help. A positive attitude makes most things easier. Let's look at the domains one by one.

NUTRITION

The phrase 'eat like a horse' is famous even outside the horse world. Horses need to eat mainly fibrous food for around 16 hours out of 24. They are not being greedy; this is just how their digestive systems evolved, and so if this need is not met, the results can be digestive discomfort, distress, boredom and frustration, and maybe serious conditions like gastric ulcers and colic. From a welfare point of view, then, we must not keep our horses short of fibrous food such as hay and haylage.

Horses' natural food is grass with tree and shrub leaves, seeds, roots and, when available, fruits. All these foods are high in fibre, which is of two kinds—digestible fibre that can be broken down into nutrients, of which they contain most in spring and autumn, and indigestible, woody fibre or lignin which acts as a bulky 'filler' and helps to physically break up other foods to enable easier access by digestive juices.

When we put horses into hard and, particularly, fast work, the practice has been traditionally to restrict their fibrous food (hay and haylage) so that the horse is not carrying too much bulk in his system which could slow him down and cause varying degrees of indigestion and even colic. Horses' main energy-providing food has traditionally been oats for their high starch (for energy) and protein content, mixed with bran (part of the outer layers of cereal grains) and chop (hay and, usually, oat or barley straw chopped short, hence the name).

Now, we still have the essential dietary basics of hay and/or haylage, they can be analysed to check their nutrient levels, including their energy content, and add concentrated feeds if necessary such as commercial mixes and 'nuts', or crushed or, preferably, rolled oats or barley, if you wish. There are also excellent bagged fibre feeds of various energy levels, which are nutritious in themselves and can have other feeds added to them. Pelleted 'balancers' are now welcomed as a palatable form of supplementary vitamins, minerals and other nutrients but few, if any, calories, fed in small amounts such as a mugful. They are ideal to mix with chopped fibre feeds instead of using larger concentrate feeds, which are not the most natural feed for horses.

The firm whose feeds you use will probably have a helpline offering free, expert advice from a qualified nutritionist who will advise you on all aspects of your horse's diet and be able to formulate a diet suitable for his type, lifestyle and work, if any, and, importantly, that will cover his innate need to be able to eat for about two-thirds of his time, so helping to take care of his physical and mental health. Contacting these advice lines is a wise move because, as qualified specialists, they will be well aware of the significant downside to underfeeding fibrous feeds. This was a regular occurrence in the past but still happens today where those in charge of horses' diets have not kept up with modern knowledge on feeding and equine health.

Water is a crucial part of a horse's diet, and it is ideal to keep it constantly available in stable and field. Obviously, it is natural for horses to drink from ground level; in this way, the process is easier for them and they drink more. A simple way of providing water in stables is in two clean buckets on the ground, scrubbed and rinsed daily,

Figure 4.2 Ike and his owner, Alexa, enjoy one of their regular hacks. A change of
scene is welcomed by most horses and ponies, especially if the route can
be varied as a matter of course, and provides variety for them.

frequently checked, and sited in different corners in case the horse does a dropping in
one (but always the same corners so he knows where to find them at night).

Automatic drinkers reduce work but need to be checked regularly for supply problems
and need some way of being emptied and cleaned out daily to prevent the formation of

slime, rinsed and refilled. Siting drinkers depend on the external supply pipe locations but also it is common for them to be fixed too high in stables, making drinking uncomfortable and off-putting so that horses may not drink enough. A horse needs at least to be able to drink with his poll lower than his withers so he can adopt a fairly natural posture of head, throat and neck for the water to travel up, then down into the stomach.

If horses drink from natural supplies such as streams and ponds, their supply should be tested regularly to check that the water is actually safe to drink. Their approaches should also be safe.

PHYSICAL ENVIRONMENT

Horses are, of course, in their later evolution, creatures of the wide open spaces but early on their genetic ancestors evolved in forests. As mentioned, it is quite likely that the inclinations of a few equines to hide, if possible, rather than run away and to be good at jumping is a throwback to their early evolutionary environment. Horses can swim but not all of them like doing so: my first horse was happy enough to go into the sea first time and ever after but would initially seriously baulk at puddles.

Horses' more closed-in, early forest environment might also explain why so many of them adapt to stables and usually come to associate them with shelter, food, water and bedding. In fact, those who do not react well to being stabled have often had an accident entering or leaving a stable, or had an unpleasant and inescapable experience

Figure 4.3 Stabling like this is very common but not very horse-friendly. Only one seems to have a rear window for an alternative view and more light, and the horses can only see each other, not socialise by means of touching.

in one, or been confined for too long. I am sure, though, that natural claustrophobia plays a part, and sheer isolation from normal contact.

Stables can range from being nothing more than dark, claustrophobic boxes with only one outlet—usually the open top door, if it *is* kept open—to a horse-friendly place a horse can feel safe, comfortable and at home. We'll centre on that positive image and see how stables can become adapted as places horses enjoy retreating to after work or exposure to bad weather. Many horses spend a long time in stables, often too long, so anything we can do to make them more horse-friendly is important.

The usually recommended size for a loose box for an average-sized riding horse of about 16 hands high/1.62m is 12 feet/3.6m square. This is actually minimal and we should aim at more space for a movement-orientated animal like a horse. He should have enough head-space to avoid banging his head if he should rear inside and to help with ventilation. Because warm, stale air rises, there is much to be said for having ridge-roof ventilators on each box, if they are separate in a conventional row, or preferably an open ridge facility with coping along the top to allow the stale, warm air out and stop the rain coming in.

The inner walls should be a little higher than a horse's back and, for safety, it is usually recommended that the space above is barred to the eaves. The horse should not be able to fit a hoof through the spaces. This way, they can see all along the row and touch and communicate with their neighbours. The feeding corner, if a manger is used (which should allow the horses to eat with their polls lower than their withers), could have a solid section on the inner partition so horses cannot see each other feeding, which can cause friction.

Providing hay and haylage can be done with various types of hay holders, from haynets to large tubs fixed in a corner, and it is best if horses have two or more positions for eating forage, for variety and to even up muscle use if using nets. It also gives an opportunity to feed different forages so the horse has variety, as in nature. The same choices should be kept and changed gradually when necessary, to maintain the health of the horse's digestive microflora populations in the hind gut, where fibrous food is mainly digested: sudden changes in feed can disrupt and kill off these populations, possibly resulting in colic, which is always a serious issue.

More than one outlook from a box is welcomed by most horses but have regard to the direction of the prevailing wind. Even so, if an open space equivalent to an open box door can be fitted in a side or back wall of a box, it can also have a toughened glass door which can be closed in bad weather.

Drainage of urine has been experimented with over the years to solve this problem, but no really satisfactory answer has been found. The actual flooring of the stable must be slip resistant in case the horse gets his hooves through the bedding. Rubber matting is popular with some and can work well if regular attention is paid to cleaning out underneath it, unless it is sealed to prevent urine getting under it. From a welfare viewpoint, I am wholly against zero-bedding systems because they reduce the time horses spend lying down, they are not particularly comfortable when they do and there is usually nowhere for the urine to go, resulting in discomfort and skin problems for the horses. Horses' favourite bedding seems to be straw, and chopped

Figure 4.4 Conventional stabling often consists of just a square box, often none too
large, with a window and door on one side only, giving a restricted view
of the outside world. The modern practice seems to be to automatically
fit anti-weaving grilles on the upper halves of the stable doors: such
grilles do not stop horses weaving because, if they want to weave, they
will do so inside the stable. The grille prevents the horse moving his
head and neck more naturally over the door, so is psychologically and
physically restricting to him. It also seems that most horses are rugged up
with a rug/blanket or lighter sheet whether they need one or not, which
can cause them a good deal of discomfort. Stable windows should open
out and down for safety, not up and in.

wheat straw is good for them and for those who have to muck it out, being easier to
manage and more economical than long straw.

Alternatives to conventional stabling are often great favourites with horses: there
is no reason why two friendly ponies cannot share a big box, or two horses in a
double space, if available. Large sheds, especially if opening on to a field or a surfaced, fenced-off area, are popular with them, having a hay/haylage supply inside
and water somewhere convenient. Ordinary field shelters can be used, and I wish
more use were made of round, hexagonal or octagonal ones because these prevent
horses being cornered by another horse and possible fights occurring inside. Their
openings should, ideally, face south in winter and north in summer, always considering the prevailing wind and weather direction.

Moveable shelters (often on skids) can be moved to prevent the entrances becoming poached and discouraging for the horses. With any shelter, the flooring needs

Figure 4.5 Horse-friendly stabling can offer such features as more than one outlook, different types of fibre feed provided so the horse has more of a choice, and the ability to touch neighbours via chat holes or bars. Of course, neighbours must be carefully chosen, but this should be done in any case because two unfriendly horses stabled next to each other will never be as content as two friends: even if they cannot socialise, they are still aware of each other's presence.

attention. Earth floors are common in large barns or sheds, but smaller shelters are often bedded down. Either way, the droppings need removing daily. Obviously, badly drained, stony or very sloping floors are not horse-friendly. If horses are housed, say overnight, in indoor or covered schools, any equipment stored in the corners must be safely and securely fenced off.

Most horses should be turned out in company for exercise much more than many are. People who claim that their horses, or those in their charge, are 'too valuable' to risk being allowed out, whether on their own or in company, show a total disregard for the horse's mental contentment which is just as important as its physical state. A deprived mental state creates poor physical health, too. Most horses, including stallions, should have many hours of exercise and freedom every day, in safe, large enough enclosures, grassed or surfaced (although almost anything is better than being stabled most of the time), and with appropriate company. It has been shown in humans that being out in nature boosts the immune system and improves both mental and physical health, and there is every reason to suppose that this applies to horses, too.

Gateways can become problematic in wet weather and just putting rubble or bedding material down in the mud is not the answer; it just makes it worse. The best choice is to have the gateway and some metres either side of it, if it is between fields, excavated out (the earth can be used elsewhere) to a depth of about 18ins/.4m. Then

Figure 4.6 Horse heaven. This is how horses want to spend most of their time, they
        have all the Three Fs—friends, freedom and foraging—and many prefer
        to be in this situation than stabled, except in real extremes of weather,
        and even then, it depends on the horse. Horses let us know when they
        want to come in, and if there is a good field shelter with hay and, of
        course, easily accessible water in the field (not an approach of deep mud
        or rough stones), many still prefer to be out together.

hardcore is laid and rammed down as a bottom layer. Next, smaller rubble on top
and rammed again, a layer of gravel and finally at least a foot of sand put on top,
rammed down, and filled up again if necessary. I have experienced this arrangement
myself and found it ideal all year round, virtually maintenance-free apart from top-
ping up with sand occasionally, and was welcomed by the horses, many of whom
hate having to splosh through wet mud to get to their paddocks in winter, and it is
better for their feet and legs. I strongly recommend it as a very worthwhile, long-
term investment.

Track systems or 'paddock paradises' are becoming more popular. In these, a
series of fenced off tracks is arranged around and through the available land and
arranged so that horses have to walk some distance to get to the water, or to the
various hay supplies if grass is not available. The tracks often have grass and shrubs
growing at the sides, and the central land is divided up into sections with a system
of gates, allowing the horses in to eat off the grass in one before the openings are
changed to allow access to a different area. A communal shed or shelter is always
made available, and horses can really live a herd-type life, often with great improvements in their temperaments, apparent enjoyment, interest and contentment, plus
physical health. They are also fitter because they have to be on the move so much.

Shelter is important to horses despite their evolution: wild and feral equines usually have many square miles over which to roam to find shelter, which most domestic horses don't. Trees in a paddock only do half the job but are better than nothing. In winter, horses gathering underneath them experience water dripping on to their backs from the leaves or branches above, which can certainly badly affect any skin exposed, and in summer, under the trees is precisely where troublesome and disease-causing insects gather. It is great to have as much natural shelter as possible, but horses, almost without exception in my long experience, prefer man-made shelters large and small because, if well managed, they are very effective year-round, and flies don't seem to like going into them in summer: hanging cut garlic or onions up in such shelters, which involves changing them every few days, does discourage insects, if you don't want to spray or treat the inside with insecticide. It is worth keeping up to date with your vet on safe insecticides to use.

Clothing can be a blessing or a curse to horses. It may seem strange to see it in a section on the environment and shelter, but it protects horses from the elements and from insects. Unfortunately, it can also cause them a good deal of long-lasting discomfort when it doesn't fit or when it isn't needed, or checked and changed often enough. For horses out all the time, at least two rugs per horse are needed in winter to keep them dry and clean.

Rugs are often worn for many hours at a time, so it is very important that they fit well for comfort and are suitable for the weather conditions as regards temperature and, for outdoor ones, their waterproof and windproof properties. They have to withstand horses getting up and down, moving in every way horses do, being rubbed up against hedges, trees and fences, other horses grabbing hold of them with their teeth, and have to stay in place if they are not to become instruments of almost torture to their wearers. Their design can vary from a light fly sheet to substantial waterproof and windproof 'portable sheds' with neck covers and tail flaps, from light summer sheets to protect stabled horses from dust and flies (which should not be necessary) to warm, winter duvets in disguise.

> Reliable studies have shown that horses prefer to be without rugs provided they are comfortable and have access to shelter and hay or haylage. There is no doubt that ill-fitting and poorly-maintained rugs can cause long-term distress.

Horses do like shelter, mainly from wind but also rain, although they can easily withstand much colder ambient temperatures than we can provided the weather is still and provided they have not been over-clipped. A horse with an extensive clip, even wearing a rug, can feel the cold significantly, and most horses are over-clipped, from a welfare point of view. For practical or appearance purposes, no horse ever needs clipping out, that is, with all his body hair removed, in my view. The worst clip to give a hunter is a hunter clip which leaves his hind quarters and loins exposed, a deprivation which horses working in winter, whatever their job, should never have inflicted on them. And clipping a horse for the sake of appearance and 'smartness' is irrelevant: it is his comfort that matters.

A point often not realised is that the advantage of the natural 'loft' of a horse's full winter coat, that is, the layer of body-warmed air between the hairs of his coat, will

Figure 4.7 Surfaced exercise areas are ideal for ensuring horses have the facility for freedom all year round, when the natural ground is just too bad to turn out on. Some people use their manèges, others have exercise areas created for this important function, often with a shelter opening on to them, with the entrance facing away from the prevailing wind. If the horses have hay, the surface is inviting enough for rolling and two or more friends are turned out together, it makes a welcome change for them, a chance to be together and is good for both mind and body.

be lost if a rug is put on top of it because the rug flattens the coat. Therefore, full-coated horses should probably not be rugged up unless there is significant wind or, especially, rain, although very fine-coated horses are excepted. If a horse has a pulled tail (plaited ones look much nicer and leave the natural hair in place), he will need a tail flap on a winter rug to protect against wind and on a fly rug to protect against insects irritating his sensitive parts.

There is no doubt that many horses gain great comfort from a good rug. So what are the features of a good rug? First, it should be horse-shaped all over, that is, it should follow the up-and-down curves of his back from withers to tail, it should have elbow and stifle darts, pleats or other shaping devices to accommodate his shoulders and hips and be shaped with either darts or a drawstring, around the back edge to reduce wind blowing up underneath it when the horse shelters, as horses do, tail to the wind, rain, snow and sleet.

As regards fit, any rug needs to come well forward of the withers, which are very susceptible to rubbing if the rug slips back on to them, and should come up round the base of the horse's neck, not rest further back on his shoulders creating pressure and friction. At the back, the rug or sheet should come right back to the root of the tail; a winter rug could well extend a little past that point. As for depth, winter rugs can extend to halfway down the forearms and gaskins, summer ones about level with the elbows. If a rug is going to be worn regularly or even daily, you will certainly need two and maybe three to provide a spare for when repairs, laundering or cleaning are necessary.

On this topic, perhaps we should all consider if our horse really needs a rug at a particular time, whether he really need clipping, at least extensively, and what rug really suits *his* requirements rather than our preferences.

Odour is a difficult subject to explore when it comes to horses. The only odour I have ever known horses to react violently and aversively to is that of pigs. Lots of horses seem afraid of pigs and I have never worked out why. I have never had the courage to visit an abattoir but have been told that horses sent to one for euthanasia react to the general smell of what can only be described as death—the blood and, perhaps, the sense of fear as stress hormones are given off by animals already present. It does seem true that horses and other animals recognise places from their past by smell when they revisit them, even starting to react while still in their transport vehicle, and they certainly recognise people from their scent, again as do other animals. From a practical viewpoint, it is wise for us to keep to our usual toiletries when around horses as they become associated with our personal body scent and changes can confuse horses.

On this aspect of scent, in this case connected with 'bonding' with humans, a friend of mine adopted an elderly horse who didn't like humans. She was doing well with him and one night forgot her coat, leaving it hanging just outside his box. The next morning, it was in his box and had clearly been lain on. She left it there every night after and he always took it into his box and slept on it. We liked to think it was because he wanted her smell to comfort him rather than that he didn't like his bedding.

I have known several horses who would not roll or lie down in their stables unless the bedding was clean, but many make a point of digging and rolling when a lot of fresh bedding is put down. This could be because they like the feel of it, or maybe want to put their own scent on it. Horses certainly identify grasses and other vegetation by smell, plus the food we give them. Once, when my horse was on box rest, I spent over an hour filling a haynet with grass for him from his normal field, but he wouldn't touch a blade of it: obviously I had collected the wrong sort.

Horses are always keen to investigate other horses' droppings when out hacking, for identification purposes presumably. I once spent a day with a troupe of circus liberty stallions for the purpose of writing an article about them and was interested to learn from their owner that one of the most important things was to train them to do a dropping before going into the ring so that the surface was not soiled for the other acts, and also to get them to urinate before going into the ring, otherwise the audience would find it either embarrassing or hilarious!

The way of achieving both was to have a designated place and to place a pile of another horse's droppings on the ground. Then the stallions were all brought along in turn, sniffed the droppings and did their own on top of them, following their 'marking' instincts. The next horse would sniff the fresh droppings and cover them in turn, and so on. The same technique worked with the urine, which was taken in a bucket and a little poured on the ground. This behaviour is common in the wild and in stallions turned out in turn, separately, in the same paddock on studs, to disguise the excreta of the horse who has gone before.

Sometimes it is necessary for a horse to change stables, some yards doing this regularly for less than good reasons and without considering the consequences. Of course, there could be good reasons such as two horses not getting on, or one needing a quieter or busier spot, but generally, once a horse seems happy in his space and especially if a friend is next door, it is best to leave him be. If change is necessary, because of horses' acute sense of smell both boxes should be thoroughly cleaned so that as little as possible of the previous horse's scent remains; obviously all the old bedding must be removed not least because of hygiene, but the floor, walls, windows and ledges and the tops of the doors should be cleaned with a safe cleaning product and, ideally, rinsed to try to remove the smell of the cleaning agent. Steam cleaning is often used to save time and probably removes old scents effectively.

Scent is obviously important to horses and is used in some experiments involving food in covered containers, and so on. Because it is important to them, we should surely let them sniff anything they find of interest, within reason, to let them fulfil their sensory needs.

Noise of any kind has some similar effects on horses as on us. Our voice can provide useful training and riding aids or cues, if used fairly. Usually, to get horses to understand a word or phrase, it has to be spoken in as identical a tone and way as possible and in the same context so that its familiarity can be linked with a particular response, initially learnt from a physical cue or signal, such as learning that the signal for walking forward is a squeeze from the rider's calves. When he has learnt that, and he is responding to a lighter and lighter squeeze (becoming classically conditioned to it), 'walk on' can start accompanying it either *immediately* before the squeeze or accompanying the start of it so that he can link, or associate, the two—associative learning.

It used to be taught that, to avoid overburdening the horse's limited mental capacity, we should use as few vocal aids as possible. Now, however, we know that horses can discriminate lots of different vocal sounds from us. In her fascinating book *Equine Education*, Dr Marthe Kiley-Worthington (*see* Recommended Reading at the end of this book) describes on pages 78 and 79 how horses can learn up to 200 words, and anyone who has seen her working with them will not doubt this.

Horses become familiar with the sounds made by other creatures, plus noises from the environment. They come to learn which car engine sound is associated with which person, that the rattling of feed buckets on a well-ordered yard means food is on the way but also that, on a less well-run yard, such as many DIY livery yards, rattling feed buckets do not necessarily mean that they individually are going to receive food. Sounds over which we have no control like aeroplanes, sirens or phones ringing might upset some horses at first, but they quickly become conditioned to them and learn that they have no consequences, so come to ignore them.

People generally do not use their voices in the right way. Horses hate shouting and screaming: it upsets some and others ignore it but in a negative, 'helpless' way. Mostly, when, for instance, competitors at all levels and in all disciplines 'praise' their horses for doing well, they thump and slap them which is meant to be an enthusiastic version of a rewarding pat but which the horse dislikes because, as explained earlier, it means in horse language 'go away' and could almost be seen as a form of abuse.

The vocal accompaniment to this is usually an erratic, screamed 'good boy oh you're wonderful you were brilliant' and so on, which the horse cannot understand but also finds offensive or even a bit scary. Also, as the horse has finished what so pleased the rider, there is no point 'rewarding' some minutes or even some seconds after he has passed the finish line, or whatever, because his brain cannot connect the reward with his action. As will be seen later, reward, or reinforcement in ES terms, has to be given within as little as one second of doing something right for a horse to connect it with what he has just done.

Our vocal quality or tone needs to be quiet, confident, calm and identical for each individual requirement or aid. Horses take great comfort from being reassured by such a voice, and words or sounds should be devised for calming horses, 'eeeeeeasy' and 'aaaaaaall right' both working well if spoken in a low, drawn out sound, if possible accompanied by firm, kind *stroking*—no patting, remember.

Predictability may be a surprising element of the Five Domains, but it is amazing how much more settled, content and outgoing horses are when kept on a familiar, predictable routine. All changes with horses should be made gradually, from daily routine to diet to new companions to new carers. When horses have a break in routine because they are, say, out competing or whatever, they soon come to know the different situations and programmes, although it helps them to have familiar, trusted handlers with them.

I once had a much-loved mare at probably the best livery yard I have ever used. I trusted the owner and her main member of staff and was quite happy that the other two pretty useless ones were banned, by my mare, from her box. One morning I arrived at the yard and could tell by the troubled heads poking out from the boxes that something was amiss. My own mare was weaving like crazy which she only did when really upset, another was tossing his head up and down, one was crib-biting, several were banging their doors and others just standing looking confused and distressed.

It transpired that the yard owner's mare had been taken out by a friend for a ride before breakfast time and the owner had delayed first feed (although not hay) until she came back because she didn't want her to work straight after a feed. She thought they should all be fed at the same time (fair enough) so had delayed breakfast till her mare came back instead of keeping to routine and feeding as normal after her mare had gone out, leaving her feed in her manger for her return.

In natural conditions, horses eat for most of their time, are prone to bouts of energetic larking about and, in some regions of the world, are subject to being preyed on at any time. So they sometimes have to gallop at top speed with a stomach and intestines full of bulky, fibrous food; yet they don't all go down with colic, otherwise the species would probably die out.

I mentioned this to the yard owner, who didn't think it relevant, then I went off to the feed room to get half of my mare's breakfast for her to eat while I groomed and readied her for her ride. She dived on it, hay notwithstanding, leaving her usual spoonful in the bottom of her manger. Out we went, walking for half an hour before trotting and cantering, then walking home to the rest of her breakfast which I had left in her manger. The other horses had only just been fed. This break in routine upset 20-odd horses, who expected their first feed as well as their hay at the usual time. My mare was fine, no colic and no more weaving.

On a visit to a police horse yard, again for the purpose of writing about them, they had a skeleton routine for the day, but horses were coming in and out at varying times from early morning to late evening, and the 30-plus horses never had a communal feed time but had hay available most of the time. Apparently, there was never any upset at the routine being broken because there wasn't one for all the horses: they all knew, though, that they would be fed at certain times when they were in, although this varied, and again when they came home. They were all in good condition and did not seem stressed out to me, and the yard manager said that colic was very rare. So in a way their lives *were* predictable though not strictly routined in the usual way.

Horses do adapt to different routines: every day may well not be the same; few horses work and live to a set pattern seven days a week, and they seem to understand the pattern of events on some days. Generally, predictability can relate to other things, like a rider's way of giving aids, routes for rides, differing hours of working in, for example, riding school horses and working in different places when at competitions. I think that routine and predictability *are* important, but perhaps the most important aspect of a working horse's life is that he knows he will be given what he needs and wants regularly and feels secure in that knowledge—if that doesn't sound too anthropomorphic.

TRAVELLING

One of the most stressing things we do to our horses is transport them. Even though they may load and unload without problems, travel well and eat during their journeys, balancing in a moving vehicle demands continuous muscle use and horses are often very tired on arrival. It has been said that one hour in transit equals two hours hard work because the muscles used are not used consistently several days a week and are not those a horse would naturally use much in any case. Even if horses are fit and well, it is a good idea, for daily trips, to ensure that they arrive at least an hour before they will be required to work and, on longer trips, it is recommended that the vehicle stops every two hours and the horses unloaded to be able to get their heads down to clear their respiratory passages of any accumulated fluids and inhaled debris. Without this, horses are at higher risk of developing systemic inflammation and 'shipping fever', a type of pneumonia infection, particularly if the vehicle is also poorly ventilated.

Today, despite ample evidence that it can be harmful to horses, most horses are travelled facing forward in the direction of travel and tethered to a ring at the side of their head: the rope is long enough to allow them to eat hay but not to get their heads down. In addition, the action of accelerating throws the horse's weight back on to his hindquarters which are usually least equipped to handle it, horses being 'forehand-heavy' by having almost two-thirds of their weight on the forehand. Braking produces the opposite thrust, which can frighten some horses who are in fear of banging their heads on the solid partition in front of them. The whole process can be very stressful and tiring, especially with a careless driver; stress hormones start circulating and can reach high levels, so the flight-or-fight response can be triggered in some horses which, combined with fatigue, makes them quite unfit for work on arrival.

It has been known for several decades that the most horse-friendly way to transport most horses is tail to the engine which can avoid or greatly reduce the occurrence of the previously mentioned significant objections to forward-facing travel. Some vehicles—horse-boxes large and small and also some trailers—are available to enable rear-face travel and easier loading designs, but, the horse world being so conservative as it is, this mode of travel is still not appreciated or used sufficiently and will not be until horse people create the demand for it, for the sake of their horses.

FEET AND FARRIERY

We all know what it's like to have sore, painful feet. They affect nearly every movement we make, we can't walk properly let alone run, they affect our balance, the comfort of our whole body because we hold and use it differently, and so our attitude to life. A farrier can make or break a horse: skilful, sensitive trimming, the willingness to make the shoe fit the foot and not vice versa, choosing the right design of shoe, his or her handling of the horse and their understanding while fitting, burning-on if hot shoeing (not too long, please), and nailing on all affect the horse's attitude to this vital aspect of horse care, his soundness and, understandably, his attitude and willingness to work.

The reason for not burning on any longer than absolutely necessary (the giveaway being clouds of sulphurous smoke) is because it dries out and shrinks the horn. The shoe is then cooled in cold water, of course, and nailed on but when the foot is subsequently exposed to wet conditions the horn naturally expands again, so the horse's shoes feel too small and tight, the horn overhangs the shoe slightly, creating discomfort and all the problems that brings. The usual reason given for prolonged burning-on is that it burns a 'bed' for the shoe, but that should not be necessary in a well-shod foot and is also a way of manipulating the foot to fit the shoe instead of the other way round.

A good in-between method is for the farrier to trim the foot as required, try the shoe and adjust it hot, cool it, try it loose against the foot again, make any amendments hot, cool it again, and when it is right, to nail it on. This does take longer and, so, costs a bit more, but it does away with the possible disadvantages of burning-on and is ideal for horses who are obviously frightened of hot shoeing. Some horses, racehorses for example, are nearly always shod cold.

If you find a really good, considerate and communicative farrier, treat him or her tenderly—ample coffee or tea, biscuits or other food, especially cream cake, I find, and (almost!) whatever else is requested are the order of the day—plus paying your bill in full on the day.

Seriously, it *is* true that, working, resting or retired, a horse is nothing without comfortable, healthy feet, shod or not. 'No foot, no horse' is a very true, old saying. For the purposes of this book, with its emphasis on welfare, I just want to discuss one misunderstood aspect of farriery—working barefoot, which has become more and more popular over the past decade or so.

WORKING BAREFOOT

Research has shown that working barefoot allows for the structures and functioning of the feet and legs to operate more naturally, with freer movement of the various bones, joints, tendons, ligaments and muscles, better blood circulation and more natural expansion and contraction of the hoof in motion. But it has to be said that, depending on their precise underfoot working surfaces, it does not, by any means, suit all horses or ponies.

Whether you use a farrier or a barefoot trimmer, horses need very gradually accustoming over many months to working barefoot and they still need sound horn quality, a pro-hoof diet and good hoof conformation, with no significant limb movement abnormalities which will create quicker, uneven and excessive wear. The feet still need expert trimming, the intervals between trims depending on the horse's work, working surface, horn quality and speed of growth. A horse needs exceptional hooves to be comfortable working shoeless on man-made, gritty, gravelly or stony surfaces; otherwise, no amount of accustoming him will help.

Probably, the breed least suited to working barefoot other than on turf or fairly good to soft ground is the Thoroughbred, noted for not having the best of feet. Any horse or pony with poor horn, flattish soles or irregular action will probably not be happy without shoes. They would be better fitted with medium to lightweight shoes, depending on their work.

It is understandable that owners will want to economise where possible on the care and management of an animal as expensive to keep as a horse, but extending shoeing/trimming visits by more than a week or so from the usual approximately six weeks can be asking for the start of trouble. Overgrowth of horn, if shod, alters the bearing surface of the foot and its working angle, clenches can rise and shoes become slightly loose until the horse is going unevenly (slightly sore on all four feet) or there is an accident due to clumsy action or a lost shoe.

The common, traditional practice of shoeing horses only in front is also not a good idea. Horses do carry most of their weight on their forehands naturally, which is why front shoes are put on to protect them, but the hind are left bare. This overlooks the fact that horses push themselves forward with mainly their hindquarters via their hind feet, so they come in for a lot of wear, friction and pressure.

Another overlooked point regarding this is that, if we ride well, we train our horse to take a little more of his weight back on to his hindquarters to lighten his forehand, promote a light bit contact and improve his strength and balance when under our weight—but where is that shifted, total weight of horse, rider and saddle, going to go? It goes on to the hindquarters, down the hind legs and ends up on those two little back feet, which have a ground surface often not as big as our hands. If the hind feet are unprotected on any demanding surface, even if of good quality and conformation, they can become painful, the horse cannot work well and may be rejected or retired as not fit for purpose. To me, it's just not worth the risk.

Hoof boots are often used on shoeless horses and ponies; some seem successful, but others are not. Boots are, of course, unavoidably somewhat cumbersome compared with shoes or nothing, and they can change a horse's action and comfort. The skin (as opposed to horn) parts of the lower leg can be rubbed sore or even raw and earth, sand, grit and water can get inside the boots, causing problems. We can try boots, by all means, but they need very careful supervision and are not an easy way out.

DENTISTRY

Long gone are the days (before vets) when old-time blacksmiths and farriers used to rasp horses' teeth. With the emergence of veterinary surgeons/veterinarians, the duty passed to them and today we have qualified equine dental technicians, their qualifications varying depending on your country.

The need for competent dental care for our horses is undeniable. Their teeth keep erupting until, in old age, only the roots are left and diet has to be gradually changed to accommodate this. Fibre still forms the basis of oldies' diets, but it must be soft, short chopped, soaked or steamed, depending on what the horse needs and can cope with. In the UK, soaked sugar beet pulp is a great favourite but also short-chopped fibre of varying nutritional values, soaked in a small-hole net and shaken into the manger, or a bucket or bowl.

Horses' top jaws are slightly wider than their lower ones so the top back teeth (their molars and premolars) slightly overlap the bottom ones, creating sharp outside edges on the upper teeth which cut the cheeks and on the inside edges of the lower teeth, cutting the tongue. This obviously means that, apart from easy eating and good care, horses who wear bridles with or without bits, and the excessively tight nosebands so in favour today, are regularly subjected to their cheeks being pressed against the edges of their teeth and their tongues being pressured in the same way, so regular smoothing with the rasp is a basic necessity, usually twice a year. Even smooth teeth, though, can cause bruising and wounds in tight bridles which apply pressure to the cheeks and jaws. (Horses can also chip and break their teeth, particularly if they are in the habit of picking up and crunching anything interesting, including stones, so cracking or breaking their teeth.)

We all know how painful and distracting even a small wound or ulcer in the mouth can be and horses are no different. Painful or just uncomfortable mouths (and feet) account for a majority of work problems, I should think. So, savings cannot be made on dentistry any more than farriery.

DO HORSES WORRY?

According to the science, horses probably are not capable of thinking forwards or backwards in time: they live mainly in the moment, so our plan to create a happy life for them must be continually ongoing. Having said that, it is plainly obvious that horses who find their work stressful, distasteful or frightening never thrive like those who enjoy theirs. It seems, then, that horses do think about things, about their lives, about, maybe, being among people who do not care about them, or not being among other horses in a naturally social way.

In my experience, horses do seem to worry, to think about their work and their lives, and I know I shall be accused of anthropomorphism for expressing this opinion. I have seen, over a long life with horses, too many of them being forced to work and live in ways quite unsuited to them, and had the good fortune to see them change hands and become able to live a more suitable life for them, do a job they can manage and seem to enjoy, and thrive in altered circumstances.

It seems unlikely, given horses' short working hours compared with ours, that their work would have that much of a detrimental effect on their outlook, their health and attitude to life, but I think it does. If they forgot it when the work was over, it could not affect them for so many hours afterwards. I have known horses in good homes but with jobs their owners insisted they do but which the horses clearly disliked, but, like many, they happened to be of a temperament that tolerated the unpleasantness, pain and fear their work caused them. Not all horses play up and object but become 'downhearted', maybe falling into learned helplessness mentioned earlier, probably a form of clinical depression. Other horses I have known in 'the wrong job' have reacted differently, developing stereotypies and looking tense and on edge all the time, despite what could be called good care and management for them as individuals.

A happy home life can make such a difference, so if a horse does not seem content and secure at home, despite appropriate care and the proximity of congenial friends and neighbours, surely thought might be given to his work. Such horses just do not thrive, do not eat so well, can be erratic in behaviour, suffer more illnesses, especially digestive ones, and even have social problems in their herd despite not being harassed, bullied or outcast.

The Five Domains of Animal Welfare have been updated in line with increasing public interest in animal welfare, and in just about every activity ethically requiring a Social Licence to Operate (SLO). Public opinion is becoming so powerful and influential worldwide and is now frequently acted upon by governments, at least in democratic countries, that no activity involving animals can rest on traditional laurels or fine-sounding statements from their human connections. Activities involving animals must be conducted with active and practical respect for those animals, those responsible for them acting on modern welfare mores, or they will be stopped, and rightly.

As an author fairly well-known worldwide for several decades, I sometimes get phone calls and emails from horse people all over the world, experienced and novice, names and 'nobodys', asking about my books, wanting advice, contacts or even just to discuss something they have read by me or someone else. This process can vary from fascinating to frustrating. I learn a lot about horsemanship worldwide in this way, though, and try to help where I can, but I learnt long ago to give only very general advice over the phone or by email. (One of my most intriguing and enjoyable phone conversations was with a very elderly lady enquiring about the pros and cons of buying a pony for her great-granddaughter. The lady turned out to have been 'roped in', as she put it, as a code breaker at Bletchley Park during the Second World War and the tales she told, very discreetly, made my toes curl.)

I get a lot of information about past and present practices and attitudes from these contacts but, almost exclusively, they are from people who genuinely care about animals—and those who don't generally get an earful. Those of us whose lives revolve around animals and want to make *their* lives happy, comfortable and fulfilling are not alone. The more of us who stand up for them, the better their lives can be. As President Joe Biden has said: 'The President is not in charge. The people are'. Public opinion is very strong, 'soft power'. We can use it for our animals' advantage.

IN A NUTSHELL

Talking mainly about the horse's 'inner Self', all-round contentment is known to significantly enhance mental and physical health, but what about a horse's spirit? It is just as true that discontent, fear, discomfort and misery trigger poor health in general.

Thinking of horses in a (w)holistic way as regards the horse's mind, body and spirit enables us to give appropriate all-round care and training that will really benefit him. It takes into account the all-round needs of a horse, which also accords with the shared ethos of classicism and ES, and which goes far beyond handling, training and riding. Ignoring horses' needs makes for substandard care which can result in poorer health, welfare and well-being. The Five Domains of Animal Welfare, updated from the previous Five Freedoms, are nutrition, physical environment, health, behavioural interactions and mental state. The importance of meeting high and appropriate standards in all these areas is stressed.

Travelling is much more tiring, stressful and wearying than we may realise, not least because most horses are travelled facing the engine. The advantages of more naturally balanced, rear-face (tail to the engine) travel are briefly explained. Farriery is also covered, including the probable misunderstandings about hot shoeing, working barefoot and shoeing only in front, none of which are panaceas for all or, indeed, most horses. Hoof boots are not always the answer and can create problems themselves. The importance of dentistry is stressed, and how and why tight bridles injure the horse's mouth is covered. The question 'Do horses worry?' is asked and considered in relation to horses doing a job they like and are suited to—or vice versa. And the author's opinion based on experience is yes, they do worry, if they have things to worry about.

5 What's Wrong With Conventional Riding?

Horse riding it seems is becoming ever more popular worldwide. Its main source of publicity is its status now as a competitive sport and not just a healthy, enjoyable hobby giving close contact with a remarkable animal. Many people keep horses only to compete, or to have them ridden in competitions by other people and this, of course, puts the emphasis mostly on being seen to do well in those events.

Unfortunately, but predictably, the desire of some competitive people to win often seems to take precedence over the well-being and welfare of the horses involved as all equestrian disciplines have upsides and downsides. Some people actually freely admit that they only keep horses as a vehicle for their competitive ambitions. Others clearly do care about their horses, but some seem to be easily led when it comes to their sources of expert help and advice as regards keeping and competing with horses. The requirements of some disciplines' ruling bodies as to the way horses should look and work also raises questions about the effects that it can have on the horses, and to do well under their rules, competitors may have to present them in ways that may not always be in the horses' best interests. This might seem to show a lack of basic knowledge of equine biomechanics, psychology and behaviour on the part of the administrative bodies—and of equine learning theory itself.

'ALL ABOUT HORSEMANSHIP'

Over the years, some in the horse world have raised objections about the techniques often applied in modern riding which to me, as a lifelong classicist and, more recently, a student, practitioner and teacher of Equitation Science, are quite different from and I think harsher than the way of riding I was taught. The UK's premier equestrian periodical, *Horse & Hound*, has now and again published items about the differences between modern and older forms of equestrianism. Most recently (as I write), it has published readers' letters pointing out the differences between 'then' and 'now', one reader ending her letter, about modern jumping practices, with the sentence 'it was all about horsemanship in those days' (around the middle of the twentieth century).

These in-print debates at least confirm the acknowledgement that things have changed and not always for the better and that there is increasing public interest in the issues raised within and without the horse world. It seems fairly obvious that we are on the way to a tipping point in public opinion. These happen when views that were once minority ideas, perhaps regarded as a bit cranky (like being kind to animals), suddenly are taken up by large sections of society, often not connected directly with the issue concerned, and become mainstream to the point at which policy and lawmakers cannot ignore them but must legislate to improve the situation. This has

DOI: 10.1201/9781003121190-5

happened several times fairly recently in the animal world, and I can see it happening again now in relation to equestrianism.

Usually, when a vet tells you something to do with your horse, you listen and act on it. It is interesting, then, that a veterinary surgeon and horseman, Dr Gerd Heuschmann, who has written four books (so far) on the downside of modern riding, one of them specifically about classical riding versus modern dressage, has caused an uproar of objections and even insults and, I hear, threats in some sections of the horse world with his incisive remarks and scientifically founded explanations of the negative aspects of much modern riding (see 'Recommended Reading' at the end of this book). The veterinary profession is associated with the highest levels of ethical practice in any profession, and I don't think anyone could argue with any hope of success with the information Dr Heuschmann lays out clearly in his books (written to be comprehensible to lay readers) on the deleterious effects some modern riding and training practices can have on horses.

ALL-ROUND WELL-BEING

The mores of classical riding are that the whole point of good and ethical horseman-ship is to get the horse to go as well, as willingly, as freely and with as much panache as a healthy, 'happy' horse would at liberty. This is what our front cover picture on this book means to illustrate. Horse people of a modern classical persuasion, and equitation scientists, also extend this attitude to the care of horses and ponies, encouraging and promoting their species-appropriate training and management (see Chapters 4 and 6).

Horses have to be strengthened with the right work to go as near-naturally as pos-sible under the weight of a rider and saddle. They also have to be trained to learn the aids/cues/signals which ask for the various (reasonable) movements they need to carry out to make them useful and safe riding horses. The movements basically involve walk, trot, canter, gallop, maybe jump, go, stop, turn the forehand, turn the hindquarters, go backwards, go sideways, and perform other movements involving various combinations and variations of those.

Only correct, progressive work will do this. No one has yet found a way of speeding up the strengthening process or of getting horses cardiovascularly fit. To make horses work without first making them strong and fit is not only not doing the job but actually, conversely, potentially injuring them by means of excessive and possibly unnatural stresses and strains, and adversely affecting their mental health, too. Unfortunately for the horses, due to the pressures of competition this often happens today, just as some restrictive and horse-*un*friendly management methods seem to be deemed necessary to get horses competing as soon as pos-sible, or to supposedly safeguard potentially valuable horses from being injured. We always need to remember those essential Three Fs—friends, freedom and foraging.

The classical mores previously described also apply in principle to Equitation Science and, indeed, to anyone who wishes to train, ride, drive and keep their horses in such a way as to promote their health, soundness, contentment and general well-being.

Figure 5.1 A modern dressage horse and rider in competition. The rider's balance
is inclined backward rather than being upright and directly balanced on
the seat-bones, where the rider could use weight aids to better effect.
The arms are straight which detracts from a sensitive bit contact, and
the horse is slightly overbent and behind the vertical, rather than the poll
being the highest point with the head on or in front of the vertical. We see
much worse examples of this common style of modern riding, and what
a lovely horse.

THINGS TO WATCH OUT FOR

Generally, aspects of riding that can cause problems for horses and, therefore, often
for riders, too, are anything that confuses, upsets, frightens or hurts horses. Some
of the main examples common in modern riding, some of which are often actually
taught as correct, or at least effective, are:

- Very firm, sustained bit pressure which can cause inescapable mouth pain.
- Firm, sustained leg and spur pressure and application, similarly.
- Aids with opposite meanings being applied at the same moment, the common-
  est ones seemingly being 'go' with the legs and 'stop' with the bit as in 'riding
  up to the bit' and 'riding forward into halt'.
- Erratically given and mistimed aids which cause confusion.
- Poor use of negative reinforcement, as when the rider keeps applying the aid
  although the horse has responded correctly (*see* Chapters 6 and 7).

Figure 5.2 Although this rider clearly wants to go left, the horse is determinedly
veering right. The rider has leant left and collapsed at the waist, the right
heel, and therefore knee, are up and the rider is spurring the horse. A better
solution would be for the rider to sit up straight and centrally on the seat-
bones, legs dropped, incline the left seat-bone and shoulder forward a little,
ask for slight left flexion and give an indirect rein turn to the left.

- Using the whip for punishment instead of solely for training and light aids.
  Punishing a horse by jabbing him in the mouth is even more reprehensible.
- Forcing the horse to tolerate being made to carry his head and neck in a short-
  ened, cramped posture or in hyperflexion.
- Repeatedly flexing the head and neck from side to side, usually while
  over-flexed.
- Manipulating the head and neck position rather than allowing it to develop pas-
  sively as the result of correct training.

Other examples are:

- Tight bridles and nosebands and wrongly positioned bits (usually too high in
  the mouth).

- Badly fitting or inappropriate saddles, girths and rugs/blankets that can cause discomfort and pain, but which can actually permanently alter a horse's way of going.
- Poor distribution and use of the rider's weight (lack of an independent, balanced seat).
- Riding horses to significant fatigue or even exhaustion, and failure to make horses physically fit for their work.
- Using the reins to maintain rider position in the saddle.

Any occurrence or practice which involves any of those previously mentioned can adversely affect a horse's mental and physical health and, depending on his temperament, promote self-defensive actions on his part. These can include such things as:

- Rearing, bucking, 'pulling' or becoming 'strong' and not responding to bit or leg cues (because the horse has become habituated or 'used' to them).
- Bolting or taking control of his own speed and direction rather than being under the 'stimulus control' of the rider's cues.
- Shying which is often prompted by something nearby which the horse is alarmed by rather than by the rider, whose corrective cues, however, he often ignores.
- Often as a last resort, rolling or lying down when mounted. It usually takes a desperate horse to do this.

Issues such as having a 'hard', unresponsive mouth and/or 'dull' sides are rider/trainer errors resulting from poor or no use of negative reinforcement: the rider does not release the aid when the horse 'obeys' but keeps up the pressure of bit or legs, or 'nags' with the legs, believing the horse will otherwise stop. These are very common in modern riding. With the correct use of ES techniques, as mentioned earlier, I find it takes about ten or 15 minutes to initially retrain such horses, and the results will be permanent so long as the rider takes the techniques and principles to heart and uses them exclusively. In the words of my classical trainer Dési Lorent: 'Don't keep asking for the salt once you've got it', meaning stop the aid *as soon as* the horse responds correctly, which is your 'thank you' to him. Both rider and horse will be relieved!

We must not, however, release/stop the aid till he does respond correctly; otherwise, we shall confuse him or reward/reinforce whatever he was doing immediately before we stopped it, even if we were only intending to have a short rest! He didn't know that. He, being a horse, linked the relief with his action and has learnt to respond that way every time he feels that aid or pressure, to stop it, so we then have our work cut out to retrain him.

EQUINE MENTAL HEALTH

The issue of poor mental health, which is becoming increasingly recognised and the subject of studies, is shown by the presence of formerly termed 'vices' or rather 'stereotypies' or stereotypical behaviours. These include crib-biting and wind-sucking, wood-chewing, box-walking, weaving, head-circling, tongue-lolling, kicking stable walls and other similar abnormal behaviours connected to general distress and various aspects of inappropriate management (*see* Chapter 4), handling, training and riding (*see* following chapters). Sometimes, but not always, these behaviours can be 'cured' when the cause is addressed, but the propensity to perform them may readily return should the conditions that triggered them return.

The condition of 'learned helplessness' is often not recognised by us. In this, a horse learns that, no matter what he does, he is helpless to improve his situation and sinks into a condition very similar to clinical depression, in the view of some experts. Again, a horse can come out of this after a period of longish rehabilitation experiencing correct care, management and riding for the horse's individual needs and inclinations, but, as with people, can return if conditions go against him again. The important factors in recovery are (a) handle, train and ride the horse using techniques

Figure 5.3 A cross-country horse in a fast gallop which seems to be, at this point, somewhat beyond him or her. The strained expression is combined with the ear position, flared nostrils and worried eyes to all indicate distress. The strong bitting arrangement adds to the horse's anxiety.

he can understand, that is, conform to equine learning theory, and (b) manage him in a very horse-friendly way—think the Five Domains and the Three Fs.

As a teacher, I have very often had a new client tell me that his or her previous instructor taught a particular practice, such as most commonly, hard and sustained bit contact combined with vigorous and equally sustained leg/spur aids, giving plausible-sounding reasons for their necessity. The most common reason seems to be that the horse needs a firm bit contact to rely on so that he can balance correctly under the weight of a rider, and sustained, firm leg contact is needed to guide him so he knows where to go and to keep him 'up to the bit'.

Of course, horses do not balance on their mouths, they do not need to be 'up to the bit' which is a human concept, merely using it, rider permitting, to receive light directions/aids, so that they are free to accomplish that jewel of good horsemanship, well-balanced self-carriage. When horses are man-handled around with hard aids which actually prevent self-carriage developing, they either become bullied into simply trying to avoid the pain, discomfort, distress and confusion of conflicting pressures/aids from bit and legs respectively, or they sink into a browbeaten state in which they try anything to avoid those pressures, if they possibly can, or if that fails, they simply learn to tolerate them. Other horses, though, 'play up' in self-defence, potentially dangerously, as described.

How Would I Like It?

I find it particularly puzzling and sad that riders often do not think through what they are being asked/told to do by a trainer or coach, or consider the reasons they are given if they query them. It must be obvious that such hard pressure is painful to sides and, particularly, mouth, so I should like to think that all clients and students would query anything which clearly could cause their horse pain or discomfort. It can be difficult to argue with a trainer, and some come over very strongly, almost to the point of bullying, putting down and humiliating anyone who disagrees with them, but surely the welfare of our horses demands that we refuse to do what we consider potentially painful, distressing or confusing to our horses, which goes for riding school mounts as well as our own.

I suggest we all imagine ourselves inside a horse-shaped body that is just as sensitive as ours and ask ourselves how we would feel if a creature of some other species were on our back and doing 'that'—whatever it is—to us. Would we like it, would we feel a bit uncomfortable, do we think it would hurt and what would we do about it? Would we come to associate being ridden with pain, discomfort, confusion and fear, and what would we do about that?

When watching other riders, *whoever and wherever they are and whatever establishment they are from*, look for the previously mentioned issue of excessive salivation in the form of froth from a horse's mouth, of a thrashing, swishing or stiffly held tail, of a tight, strained facial expression, of horses not going straight, and squirming and paddling into halt and rein-back with straddled hind legs, and of a horse showing a neck that is shortened compared with the rest of his body (looking out of proportion), perhaps with a BTV (behind the vertical) position of the head and probably accompanied by a poll that is not the highest point of his outline.

Artificial, 'showy' action, mainly with the forelegs, which Nuno Oliveira described as a man 'shooting his cuffs', can be another sign of enforced training techniques: the cannon bones of the fore and hind legs in, for example, trot should be at equal angles of slope rather than the horse appearing to be performing a version of Spanish Trot. Horses have been bred for flashy action for decades now, but it is not a good look to those who like horses to be allowed to be horses, and because their bones and soft tissues are no stronger than those of their predecessors, they are more likely to be injured due to excessive movement. The fact that some horses are bred with this action does not lessen the prospect of injuries.

The opposite of the seemingly rounded posture is the base of the neck dipped in front of the withers, the hollow topline, hind legs out behind, head and neck concave and nose poking out stiffly. The facial expression may show wide, frightened eyes and ears stiffly held back in pain and anger, or forward in fear: this is an accompaniment to the flight-or-fight response which is instinctive to frightened horses.

More to Look Out for

The previously mentioned are common errors of locomotion today which must be uncomfortable for the horse and potentially injurious and worrying to him. In disciplines where hard, physical effort such as galloping and jumping are needed, watch for these signs:

- Horses wavering in their gallop as if exhausted.
- Excessive sweating in cold or cool weather.
- Sustained, widely flaring nostrils as the horse tries to get air into his lungs (because horses cannot breathe through their mouths), and hard, rapid breathing.
- Horses refusing to jump or running out in front of obstacles (which may be due to over-facing or associated pain, fear, discomfort or fatigue).

Obviously, these are all signs that something is not right with the horse, his fitness, his training, the rider—or that the discipline is not appropriate for that horse/rider combination.

A Secure Knowledge Base

One way to protect our horses and ourselves is to give ourselves the foundation of a good working knowledge of everything we need to know to look after, handle, ride and train our horses, without relying totally on other people to breach this possible gap. I think this book's sources of further help and information at the end, in the 'For Your Information' and 'Recommended Reading' sections, will set you off on the search for the right kind of knowledge. For books on veterinary matters and first aid, ask your vet to suggest one or two: vet books in particular go out of date quickly, so certainly update yours every five years.

It may seem futile in this day and age to recommend readers not to rely on free advice from social media or the Internet. Keeping a horse properly is expensive and

Figure 5.4 This is a general example of the type of the modern jumping position
commonly seen. The main point to make is the restriction of the horse's
head and neck because the rider is supporting his or her weight on the
horse's neck while holding the reins short and tight. Horses need free
use of their head-and-neck balancing pole to jump using natural muscle
function; otherwise, they have to use their muscles inappropriately and
soft tissue injuries can occur, especially over time. The horse will be
pushing against the bit in an effort to get his head forward, so will possibly
be experiencing mouth pain, as well as his difficulties in trying to jump.
The illustration in Chapter 10 is a good example of a comparatively
effortless jumping position for horse and rider.

it is understandable that we do not want to spend money unnecessarily, but in this
book, we are thinking of horse welfare as part of the whole ethos of classicism and
Equitation Science, as well as training and riding. In veterinary matters, feeding,
paying for services such as livery, farriery, dental work, physical therapies and so on,
cutting financial corners can be dangerous. There is no getting away from it.

Professional advice and service is almost always the best way to go for our horses'
sakes, but we can and, I believe, should teach ourselves as best we can to understand
at least the basics, and preferably more, of everything that affects our horses. We
can take advice from qualified, registered professionals, attend actual and virtual

courses, lecture-demonstrations, webinars and conferences run by such people, read books detailing the correct theories and practices, many of which now have accompanying DVDs, websites and other add-ons, and back up our own competence with the right self-learning, which can be a lifelong process. It is never time or money wasted and always reassuring and rewarding.

IN A NUTSHELL

As horse-riding increases in popularity, competition has become the reason why many people keep horses; sometimes this can mean that the demands of competition take precedence over the horses' welfare and the horse–rider relationship.

Riding techniques have changed over recent decades, and many believe them to be harsher than before. There is increasing discontent among the general public about current practices, and public opinion could trigger legislation concerning equestrianism, supported by some veterinary and professional experts. Classicism and Equitation Science are both strongly in favour of effective, humane and species-friendly principles being applied in horse training and management.

The best way to learn what is good for our horses and what is not, and to develop our own common horse sense, is by giving ourselves a sound, working knowledge of all aspects of training and managing horses, putting ourselves in the horse's place particularly and being able to critique the performance, actions and statements of other riders and teachers, whoever they are.

Horses need to be strengthened and made fit in order to work and develop safely. A horse also needs to be kept happy in his daily life via appropriate, horse-friendly management if he is to thrive and avoid physical and mental health disorders as much as possible.

6 The Equitation Science 'First Principles of Training'

Equitation Science (ES) is not a method or system in itself. It is simply the science involved in horse training, management and riding. The International Society for Equitation Science presents ten first principles of ES training, taking fully into account the kind of animals horses are, and, crucially, how they think and learn—which is not the same as humans. These principles have been researched, investigated and shown to be reliable, and it is hoped that more and more of those involved in educating horses and horse people will study and use them in their work.

They form a framework and working pattern we can trust, through which we can understand horse training and apply it to *any* system or method of horse training. They are detailed on the website of the International Society for Equitation Science (*see* 'For Your Information' at the end of this book) and were made readily available to the general equestrian public in a free webinar held on the 11 June 2020 by Dr Andrew McLean of Equitation Science International, Australia (*see* 'For Your Information'), who is one of the leading figures in ES today.

ES involves learning theory, ethology and cognition, biomechanics, psychology and sports science. It is a set of principles anyone involved with horses can follow to achieve best practice in their dealings with horses and which help to ensure equine welfare and well-being. In common with true classical riding, ES aims to establish immediate responses from horses to aids/cues involving only slight pressure, known as light aids; a horse who does this is sometimes called a 'light ride'. Lightness is accompanied by even rhythm of the gaits and maintaining the rider's required line, which horses are trained to self-maintain along with balance and tempo (speed), a state called in both classical riding and ES 'self-carriage'. A horse going in 'self-carriage on the weight of the rein' goes *voluntarily*, not compelled by his rider, with his weight shifted slightly back towards his hindquarters, spine slightly arched but vertically flexible according to his gait, his dock swinging, his forehand raised a little and his head and neck naturally stretched up and forward in an arched posture, his head being carried so that his nasal planum—the flat bone down the front of his face—is carried just forward of or on an imaginary vertical line dropped from his forehead.

It is important to realise that this posture is naturally developed and assumed by the horse in response to the strengthening and balancing work he undergoes in a correct training/schooling programme, all of which is discussed later in this book. It is absolutely wrong for the rider to force it by pulling in, shortening and fixing the neck and head via the bit as is so often done in conventional modern riding. That will not be the 'proof of the pudding' because it does not happen naturally as the result of correct work. Horses coerced in that way show clear signs of discomfort, distress,

DOI: 10.1201/9781003121190-6

anxiety and pain in the mouth, the muscles and soft tissues of the forehand and the back in particular, plus inordinate stresses on the hind legs, especially the hock joints.

A vet told me back in the 80s that dressage horses had started to sustain more injuries of the hindquarters and legs than jumping horses. In his opinion, this was because of the 'new way of riding' that forced horses into an artificial gait. Quite thought-provoking at the time.

Here is the gist of the Equitation Science First Principles of Training, plus my own comments:

1 REGARD FOR RIDER AND HORSE SAFETY

We must acknowledge that the size, strength and easily startled nature of horses make them risky to be with. We need to avoid triggering aggressive-defensive behaviours like bucking, rearing, biting and kicking and recognise horses' dangerous parts such as teeth, hooves and hindquarters. Remember that if we are directly behind a horse, he cannot see us.

We need to maintain a safe environment for horses and use and store tools and equipment safely. Horses are learning all the time, so we must make a habit of using correct techniques properly and *consistently* to avoid confusing them. We must avoid using methods, equipment and tack that cause discomfort, pain, distress or fear, and injury to horses. It is obvious that, for the safety of both horse and human, the two need to be well-matched.

When we start to study and use ES, we soon realise that the timing of applying and stopping our aids/cues is crucial to the horse's learning, as is the need to always apply them identically. Ineffectiveness leads to inconsistency which leads to confusion and possible dangerous, self-defensive behaviour.

2 REGARD FOR THE NATURE OF HORSES

I hope the points made in this book will clarify the type of animal we are dealing with. As a plains-living, grazing prey animal, certain aspects of horses' behaviour are hard-wired. The webinar mentioned earlier emphasised the Three Fs—friends, freedom and foraging—three necessities most domestic horses are very short of, as described in Chapter 1.

Horses cannot help needing to forage (search around for food) for around two-thirds to three-quarters of their time, to be able to touch their 'preferred associates' (friends and family) whether that is mutual grooming or just grazing shoulder to shoulder, and they need the mental reassurance and physical benefits of freedom, as much freedom as possible with friends and access to food, water and shelter.

Sadly, I would say that most horses do not have access to facilities that will fulfil these basic needs on a year-round basis, or at all in some cases. Deprivation of any of them (particularly foraging) can result in stereotypical behaviours—what used to be called 'vices', as though the horse were at fault rather than his management.

The bad effects of conventional weaning have already been mentioned, but separation can also greatly adversely affect mature horses, too. Stress of any kind can trigger gastric ulcers in many species and separation, even just an inability to touch others, is very stressful to them, but so is being forcibly placed near an enemy, either stabled, in transport or even out riding.

Avoidance of general management and handling practices which are horse-*un*friendly is also necessary. Such practices as over-clipping, over-rugging, removal of whiskers (unlawful in several countries), hogging the mane (my personal view), painful methods of restraint such as ear-twitching, harsh bits and spurs (which betray incorrect training) and similar unkind practices do not show a regard for horse well-being.

Signs of pain need to be spotted and accurately acknowledged, and I recommend readers to source the work of Dr Sue Dyson in this regard, in particular for the UK's Saddle Research Trust, for which she has developed an educational equine ethogram to help us all recognise even subtle signs of various emotions. (Refer back to Chapter 1 for some pointers.)

It is surprising how many established systems and methods of equitation (which involves handling as well as riding) are still based on the old-fashioned and inappropriate idea of dominating horses. It can be tricky deciding on the difference between dominance and exploitation: all training is exploitation of the horse, but there is little doubt that many horses enjoy their time and activities with humans provided we really do place their welfare and well-being first. Horses do not regard us as their 'bosses' but as part of their lives and society. They are known now not to have an all-powerful herd leader: their apparent 'place' in the 'hierarchy' is resource-based, one horse maybe having first choice of grazing, another of certain company and another of water sources.

Some horseplay can look really rough to us but, basically, horses are gentle with each other. They are also perfectly able to tell from the feel of our touch whether we are confident, afraid, considerate, kind, violent and whatever else. Touch is *very* important to horses, so I think we need to think 'kindness' and 'love' or 'affection' when we stroke them. I have long urged my clients not to pat, and certainly not to mindlessly slap and thump their horses, even though most people think of these as enthusiastic praise. Short, sharp, unpleasant sensations like those are equivalent in horse language to nipping, biting and kicking, which mean, in terms they well recognise, 'go away'—not what we intend here. We should use a firm rub, stroke or scratch on the sides of the withers or upper shoulder area, as horses do when they mutual groom; this is known to release feel-good hormones in the horse and to lower the heart rate.

All in all, our posture (upright but relaxed), the way we touch horses (confident but kind—and clear) and the way we move (smoothness, stillness and a fairly slow speed in general) reassure and settle horses. Of course, we should not shout and scream around horses, and music in their environment should be limited, if played at all, to no more than an hour at a time, quietly. Horses are known to prefer calming music such as some classical music, ballads, easy-listening music and also uplifting military music, not anything loud, raucous or 'screeching'.

3 REGARD FOR HORSES' MENTAL AND SENSORY ABILITIES

It is almost ubiquitous, and perhaps natural, for us humans to assume that other animals think and learn the way we do. In the horse world, it is common to hear someone say something like: 'He's always doing that. He knows it's wrong', or to say: 'He's really stupid. He always shies just here but there's nothing to shy at'. Even worse is when a rider decides a horse needs punishing for something or 'teaching a lesson', organises their reins in one hand and proceeds, several seconds after some perceived misdemeanour, to whip their horse hard with the other.

We need to learn not to over- or under-estimate a horse's mental capabilities as doing so can have serious welfare implications. Horses live very much in the moment: it seems that they cannot think in the past or ahead. They do, though, have unbelievable memories and seem never to forget anything. If a horse has shied and run off at a plastic bag in a hedge once, he will connect that spot with a frightening episode and expect the bag to be there every time and will shy because it is an evolutionary safety technique.

His rider can avoid the stimulus of the horse's associative memory by, say, performing shoulder-in past the spot, bent in the opposite direction. The horse will still be able to see the place with his outside eye but shoulder-in, which can be very effective in such cases, gives him something to think about and puts him under the rider's control (or rather that of the aids/cues the rider is applying—stimulus control) instead of that of his past association. Shying, incidentally, is a problem with turning, so revise his response to turning aids, dealt with later.

On a group trip to visit the well-known classicist, the late Sylvia Stanier (who, however, had probably experienced just about every kind of horsemanship on the planet), she had a groom bring into her arena an old Thoroughbred who had returned to her for retirement, having been trained conventionally by her when he was a 2-year-old. She explained, though, that she had used a very specific method of lungeing with him, involving vocal cues and body language, and it would be interesting to see if he remembered what he had learnt with her some 20 years earlier, knowing that no one else would meanwhile have lunged him in that way. He immediately slotted back into her system and lunged perfectly and with apparent enjoyment. This horse, incidentally, had the celebrated racehorse St Simon a few generations back in his pedigree and was the absolute image of his photographs—but unfortunately had not inherited his speed!

The issue of a horse being 'taught a lesson' is sadly common. Causing a horse pain means he associates it with whatever he was doing immediately before it happened, or with a particular place or person. Applying punishment (generally not recommended) needs to be done as the horse is doing whatever is not required, otherwise the horse mentally cannot link the two and the punishment amounts to base brutality. Similarly with reward: giving the horse something good such as a treat, a rub on the withers or a vocal reward such as 'good boy' must be given during or immediately after the required action so that the horse's brain can associate the two.

Horses, as prey animals, are extremely observant and learn well by associating events and objects, whether we want them to or not. We also have to make allowances for their different sight and hearing senses, as explained earlier. Horses can hear sounds higher than we can and see better than we do in the dark. Because of these differences, we need to remember that they might become hyped up due to something outside our perceptual abilities.

They also scan around with their attention and tend, even when grazing, not to concentrate on one thing or activity for very long. Training sessions need to be kept fairly short overall (half an hour is ample) and give them frequent brain-breaks during learning. When learning a new thing, try to get a good approximation first time, then try to improve three or four more times, then stop and rest or just walk around, all on a *free* rein, a necessity some people find so hard to do. Try the lesson again, then another break, then a final time and do something quite different or stop training for the day. It's fine to train again the next day for a short time, when you will find that the horse isn't quite where you finished the day before but will be there on your next try. Short sessions fairly often enhance retention and understanding and ensure the horse doesn't start feeling harassed or bored.

Watch a horse closely when he is working or learning, either with you or someone else. Although horses are good at disguising pain and feeling unwell because, as prey animals, they soon learn that predators target the weak or sick, they do have bodily and especially facial and head expressions and postures (described earlier) which tell us how they are really feeling, so we can act accordingly.

4 REGARD FOR CURRENT EMOTIONAL STATES

Most of us know that an excited or frightened horse learns nothing that we are trying to teach him, only to be wary or frightened of whatever is worrying or scaring him, whether it's us or something in the vicinity, visible or otherwise. Aroused horses produce adrenaline, which stimulates action (the flight-or-fight response), and cortisol as a response to distress. In training, we need the horse alert/aroused enough to be paying attention and taking things in: if *too* calm, he won't be inspired to learn and if too excited, he won't be able to concentrate on his lesson.

We need to avoid the use of anything that will hurt or distress the horse, such as the modern techniques of applying unremitting pressure with legs, even spurs, and, particularly, bits and tight nosebands, plus using the whip for punishment. We need to be confident, calm and patient and, above all, consistent. Horses cannot learn if our aids/cues are different for the same requirement. We must use one specific aid for one specific movement: the horse cannot be expected to read our minds and know, for instance, that pressure on both sides of his mouth from the bit might mean either 'slow down/stop/go backwards' or 'raise/lower your head'. This sort of treatment can either arouse horses or put them into a state of dullness depending on their temperament.

Rewards such as treats, rubs (not pats) or vocal praise (which must be identical each time, not chatter and not in any loud or harsh tone) must be given consistently immediately after or accompanying the correct response so that the horse can form an association between them. Help the horse to be interested and calm by keeping stints short and by stroking and vocal calming, such as a low, long-drawn-out word such as 'eeeasy' or 'aaaall right'. Do not use 'good boy' to calm him, for obvious reasons! Do not give food when the horse is excited or frightened as he will form that association.

Training a horse to lower his head (so that the poll is at least level with the withers or below them) is a good way to calm him, and easy to teach from the ground. He must do this himself on command: it must not be achieved with gadgets or other gear as the association of free will and voluntary response will not then happen.

Have a treat in one hand, let him sniff it, then lower your hand. *As soon as* his head starts following your hand, say 'head down' (and in the identical way ever after) and give him the treat—no teasing. The treat is your 'primary reinforcer' or main reward—it acts as his motivation to repeat this action.

Giving food treats from the saddle is not good because you cannot give them quickly enough for the horse to associate them with what he has just done (the same goes, of course, for lungeing and long-reining, if used). So you can combine giving him the treat with 'good boy' or whatever you choose, but you must get the timing right and keep the words the same every time. Then when you are riding, he will come to associate 'good boy' with whatever you praised him for (so make sure you praise the right move). 'Good boy' is your secondary reinforcer. You can, of course, rub his withers straight after saying 'good boy' when riding. During in-hand work, you can say 'good boy', then give him the treat, which works even better.

Of course, when riding an excellent way of reinforcing/rewarding a good movement or productive stint of training is to let your horse have a completely free rein. It is surprising how many people cannot keep their hands still on the rein, so constantly irritating and confusing their horse by inadvertently giving what can be taken by him to be confusing aids/cues. Many also are unwilling to let the rein right out so that they are holding the buckle end. These skills must be learnt, maybe initially in the presence of one's trainer, or even on a horse simulator, for the horse's peace of mind and as an invaluable way of calming and rewarding him: standing a horse on a fully free rein with head down is a crucial technique to learn. When the head is lowered *voluntarily*, the horse's blood pressure and heart rate drop, relaxing hormones circulate and the horse relaxes naturally.

My classical trainer of the 80s, Dési Lorent, had a foolproof way—at least, I never knew it to fail—of calming down his horses, who might get upset if a new rider, or one just progressing to the horse's level, wasn't coping very well with Dési's rigorous instructions, and the horse was becoming fraught. We were taught to completely stop all aids, then place our left hand up behind the horse's left ear and give one long, firm stroke right down to the withers on the left side of his crest. They switched off like magic and sauntered to the gallery in the school for a treat. You can train this 'Stroke of Magic' from the saddle by simply stopping all aids, sitting with completely relaxed seat and legs and giving the horse a long, ideally free rein and the long stroke, all at the same time. It is invaluable because a stressed horse, and maybe rider, learn nothing good.

Adrenaline levels lower fairly quickly, but cortisol circulating around creates longer-term, chronic stress, from one bad but significant experience to changing homes, losing a favoured companion or just moving stables on the same yard. Chronic stress, depending on the exact circumstances, is really bad for a horse and can trigger gastric ulcers and other issues, but it can be relieved by paying close attention to the horse's state of relaxation and contentment, so everything should be done to help him as regards his daily living circumstances and treatment.

5 CORRECT USE OF HABITUATION/DESENSITISATION AND CALMING METHODS

There are several ways of accustoming horses to objects and situations, which are covered in ES textbooks, but here are the basics:

Systematic desensitisation: This is probably the most usual way to familiarise a horse with something. We gradually approach the object and, if possible, touch it ourselves, stand by it, maybe lean on it and let the horse examine it voluntarily, calming him meanwhile. Again, if possible, we can bring the object very slowly closer to the horse but only once he has calmed down. (Traditional police horse training is to turn the horse away from a frightening object or situation just *before* he is going to object, in his trainer's estimation.)

Overshadowing: The trainer/handler gains control of the horse's leg movements by getting him to step forward and back and repeating this till he is still a little concerned but not frightened. An object can be gradually brought nearer *provided* it no longer alarms the horse. Stroking the upper shoulder/withers helps calm and reassure the horse during this process.

Counter-conditioning: This is when we get the horse to associate something scary with something pleasant, such as wither scratching or treats. I find that allowing a horse to graze near the frightening thing or situation also works well, perhaps partly because the horse's head is down.

Approach conditioning: As the expression suggests, this means that we teach the horse to approach scary things by making use of his natural curiosity. A second person can be 'in charge' of the item and gradually moves it further and further away from the horse, whose trainer encourages him to approach, even 'chase', it. If the object stops, and the horse becomes worried, the object can be moved away again and the horse encouraged once more to investigate it.

Differential reinforcement: In this, we ignore unwanted behaviour but reward a desirable alternative. This can be difficult for people to accept, particularly where biting and kicking are concerned, but the idea is to use both negative and positive reinforcement (explained in Chapter 7) so that a good alternative (i.e. not biting) is rewarded but biting is ignored, rather than punished which is the usual response. If, for instance, a horse kicks out when a leg is touched, brushed or fitted with boots or bandages, we must keep contact with the leg till he stops, then immediately positively reinforce (reward) him. If a second person can encourage the horse's head down, this will help.

Stimulus blending: A stimulus is anything that causes a response, good or bad, personally applied or environmental. Blending stimuli involves applying something to which the horse is used and does not object to with a similar one that frightens him. The usual, excellent, example given to explain this is the blending of spray hosing with aerosol spraying, which many horses are frightened of. The water spray is applied first and then the aerosol is used a little away from it and gradually applied as well, the water is gradually removed and the horse becomes used (habituated) to the aerosol. During this process, a person at the horse's head giving him treats or rubs when he does *not* object to the aerosol is helpful. Saying 'good boy' to calm him while he is objecting is obviously undoing all your good work (praising him for objecting) and confuses him.

Flooding: This is a technique that literally floods or overloads a subject with a procedure or forced proximity to something frightening or disgusting to them while preventing them from escaping. I find when talking to equine scientists that they seem generally to avoid the use of the word 'cruel' but, in my view, that is what flooding is. Responsible ES authorities neither use nor recommend it as the risk of physical injury is great and that of possibly permanent mental damage, such as learned helplessness or depression, is highly likely. We hear a lot about post-traumatic stress disorder in humans, clinical (practical) depression and other mental health disorders and know that they can remain with people for life. In my experience, this can just as likely apply to horses and other animals. Flooding, including, in my view, for the purpose of scientific experiments, should never be used.

Response prevention: This is, rather obviously, when an animal is prevented from responding to things that happen to him, such as in flooding. Such situations should be avoided, but response prevention can happen in everyday equestrian life, not least in conventional riding in which opposing aids, signals or cues are applied to the extent that the horse is extremely confused and cannot respond to either cue. Even something as common as an over-tight noseband, preventing a horse swallowing his own saliva or opening his mouth to relieve his pain, could come in this category.

6 CORRECT USE OF OPERANT CONDITIONING (INSTRUMENTAL CONDITIONING)

This is the process through which a horse learns from the consequences of his reactions to, say, an aid or cue how to respond to it in future. The correct use of operant conditioning was described by Andrew McLean during the webinar mentioned as 'the heartland of training'. Using it wrongly can lead to serious behavioural problems that are not the horse's fault. They can show as aggression, escape and apathy and can adversely affect a horse's welfare and well-being.

Both *negative* and *positive reinforcement* (reward) will make our desired response more likely if they occur *immediately* the moment the horse performs that response: therefore, our timing is crucial. The likelihood of that cue/aid stimulating that desired response in the future is thereby increased, whereas positive or negative *punishment* will make its future performance less likely.

Our aids must be removed the instant the horse starts to respond as we wish so he can link the two. The removal is the horse's reward or lightbulb moment, although stroking and verbal praise can also accompany them or come a split second later, which is called *combined reinforcement.*

Negative punishment is removing something the horse wants (such as food) until, for instance, he stops weaving, which will show him that he is only fed when he is still. *Positive punishment* can be illustrated by electric fencing because the punishment has to be instantaneous for it to teach the horse what not to do (go near the tape). Most people find it difficult to apply so quickly when handling or riding, so it is best avoided, not least because it discourages learning and the horse can form unpleasant associations with the process or the person applying it (like thrashing a horse while holding him in front of an obstacle he has just refused) rather than with some 'crime' he committed before the punishment.

7 CORRECT USE OF CLASSICAL CONDITIONING

Classical conditioning can be described as 'getting used to' a stimulus whether an aid/cue or, say, a sound in the environment. The most obvious is the rattling of buckets at feed time: the noise precedes the horse's receiving his feed. Sound can also predict unpleasant events as well, such as clippers being switched on, or a particular engine sound heralding the arrival of the vet. In training, we must always start by giving a light version of our aid then increasing its pressure or intensity if our desired response does not happen. Then, when the horse experiences the light aid, he will respond rather than experience the 'stronger' aid that he knows is coming. In this way, the horse's responses can be obtained from subtler versions of the same aid.

When the light signal always produces the right response, we can use classical conditioning to introduce different aids/cues for the same response, such as seat aids or voice aids, by giving them immediately before the already-learned light aid. The acquisition of lightness in all aids is part of the classical and ES ethos, so it must always be used before increasing intensity or pressure: an ethical and good trainer strives for lightness in all equitation, knowing that heavy aids (and incorrect negative and positive reinforcement) can cause a horse confusion and distress, and so jeopardise his welfare. Unintentional use of aids such as random bit and leg pressures can greatly confuse horses, especially young or green ones: although they can become inured to them, we must be as careful as possible not to give unintended signals. Everything we do to a horse teaches him something, good or bad, wanted or unwanted.

8 CORRECT USE OF SHAPING

'Shaping' is the term used to indicate progressive improvements towards a desired response. A training goal is broken down into single, easy steps and each correctly performed step rewarded/reinforced towards the final result, which itself can be quite a complex movement. We should make learning easy for the horse and not overload him with new things, but only change one aspect of lessons at a time, such as place (ideally quiet with no distractions), response sought and new cue/aid. In this way, horses learn quicker and better.

The place in which we first start to train a new thing should be familiar to the horse, no distracting noise, no deep bedding or anything that will put his mind on food, such as straw bedding or hay within reach, and off training, and, if possible, a horse he is friendly with nearby: this is known to support a horse's calmness and feeling of security. Do not change the place until the horse is reliable in his responses to whatever you are training him to do. Let the horse take as much time as he needs to learn so there is no tension coming from the trainer to meet a deadline. The horse needs to feel safe and be calm.

Poor shaping with an irrational progression and rushed training cause tension, lack of understanding and, therefore, confusion. The ES training/shaping scale is excellent (*see* Chapter 7), whereas the conventional one of Rhythm, Suppleness/looseness, Contact, Impulsion, Straightness and Collection does not, in my experience, state qualities to be achieved in the correct order and omits others. It does not even mention the breakthrough moment of the horse's achieving a basic attempt at the desired response (in-hand or under saddle) and so triggering the correct use of negative reinforcement, around which most good, modern riding revolves. The topic of Contact still mentions the horse's 'acceptance' of or 'submission' to the bit which has a ring of domination to it, and long before the development of ES, I was explaining to my clients why we should put Straightness before Impulsion—the last thing we want is a crooked (therefore not straight) horse going with impulsion and developing the muscles and mentality of an incorrect and potentially harmful way of going.

9 CORRECT USE OF SIGNALS/CUES/AIDS

A golden rule of aid/cue/signal application is that we give one cue for one action, or desired response. The horse cannot be expected to be able to read our minds and differentiate which of more than one move we want from one identical aid. This is a recipe for instant confusion. It is crucial that the horse can work out exactly what action on his part will remove what level and type of pressure—that is what aid application is all about.

Although we call horses 'willing to please' and 'obedient', what is really happening is that horses are moving their bodies in a particular way to avoid (stop) the slightly irritating pressure of our aids, because we have taught them this via negative reinforcement. The aids can become so light, and the horse so conditioned to them, that they become a 'conversation' between horse and rider; mostly, the rider asks for something and the horse answers, but sometimes the reverse is the case and the rider has to investigate the issue.

It is also important that our aids are as individually identical as possible each time they are applied according to what we are asking for, so that the horse can discriminate between them—leg aids in the same place for a specific movement, bit aids with the same feel and pressure for a specific response, the same for voice and seat aids, and for our body language when working or handling horses on the ground. Horses are extremely perceptive and will spot and perhaps become anxious about any discrepancy or differences.

One of the major faults with modern, conventional riding is that opposing aids are often given at the same moment, the most common example of this being a leg aid for 'go' and a bit aid for 'slow/stop/rein back' (all of those three use the same muscle groups so the same bit aid is appropriate in this case). This pairing of opposing aids is commonly used for, for example, 'riding forward into halt' and 'riding your horse up to the bit'. Both of these can cause confusion because the horse has learnt that a squeeze from the legs means 'go' and pressure on both sides of his mouth means 'slow/stop/go backwards'—and neither instruction is necessary or appropriate. If we maintain a light but present, in-touch contact, that is enough: if our purpose for 'riding up to the bit' is to create more energy or impulsion, more bit pressure interferes with that. And we get a much clearer, straighter halt if we apply a 'slow' then 'halt' aid and stop moving our seat with the horse's back movements. No leg aid, of course, because we don't want to go forward.

No creature on earth, not even humans, can go forward and backwards at the same time so confusion reigns and welfare can be compromised. Aids in conventional riding are almost ubiquitously applied with consistently very firm pressure and sustained pretty much all the time the horse is being ridden which, again, can certainly compromise both physical and mental welfare. Any aids have to be given with a firmer/harder pressure than the sustained one if the horse is to identify the difference; lightness, therefore, is unattainable and both performance and welfare suffer.

A basic knowledge of how horses move in their various gaits is needed for an aid to be applied at the right time. There is no point giving an aid for a particular movement if the leg that starts it is rooted to the ground in its weight-bearing phase: giving our request as soon as the appropriate hoof lifts means the horse has plenty of time to understand and cooperate. Horses think like lightning.

- When a hoof is on the ground and still, the leg is said to be in the '*stance phase*'.
- When a hoof is in the air, the leg is in the '*swing phase*'.
- When a leg is moving forward, it is '*protracting*'.
- When a leg is moving backward, it is '*retracting*'.
- When a leg moves away from the body, it is '*abducting*'.
- When a leg moves towards the body, it is '*adducting*'.

The three gaits used most of the time in English-style riding, as it is known, are walk, trot and canter. There is also rein-back, gallop and a point between canter and gallop commonly known as a three-quarter gallop often used in fitness training. There are types of walk, trot and canter involving slower or faster speeds and rhythms and specific names.

The walk is a symmetrical, four-beat gait, always with one foot on the ground, so there is no moment of suspension with the horse in the air. The sequence of footfalls, for instance, is left fore, right hind, right fore, left hind. The horse's head and neck move forward and back in time with his rhythm, and to left and right when the left and right fore hooves respectively are landing, so the rider must follow this swing with the hands or by opening and closing the fingers, depending on the type of walk being performed. This is explained in detail later in this book.

The trot is a symmetrical, two-beat, diagonal gait, each beat divided by a moment of suspension. The left fore and right hind form one diagonal and the right fore and left hind the other. It has been customary for many years now for the rider to sit in the saddle, when on the right rein, when the right fore is in its swing phase and rise when it is retracting on the ground, and vice versa, although we oldies were taught the opposite way when young. Now, it is generally thought that asking the outside hind to bear weight *and* push the horse forward on the outside track of a curve (which is longer than the inside track) is asking too much: therefore, the weight-bearing hind is now the inside, with the added advantage, some believe, of making it easier for the horse to balance. Horses never sustain bends, when moving naturally, for as long as we ask them to do, and under weight as well, so this argument is reasonable.

In a sound horse, the trot is normally a stable, equally balanced gait in which the head and neck remain quite still and, therefore, so must the rider's hands, keeping a consistent, in-touch contact—which may not be so simple because of the 'bouncing' gait from one diagonal to the other, with the rider either rising to the trot or struggling to master sitting trot (although the solution to this is a treat to come!).

The canter is an asymmetrical, three-beat gait, the footfalls on the left rein being right hind (the 'initiating hind'), left hind and right fore together, and left fore, then a moment of suspension. The head and neck move gently back and forth in canter so the rider must accommodate this with the fingers. For most of the canter stride, the horse's back is slightly positioned with the side of the back on the side of the leading foreleg slightly forward of the other side. This means that in left canter, say, the left side of the horse's back is slightly forward most of the stride, so the rider should place the left seat-bone, and shoulder above it, slightly forward to accord with this positioning. It is often taught that the *outside* shoulder (the right, in our example) should be forward to 'help and guide' the horse round the curve, if, indeed he is on a curve, but this twists the rider's torso and is less comfortable for both horse and rider.

The three-quarter gallop is more a moderate gallop than a fast canter, just being four beats. The rider adopts a slightly forward seat with shoulders ideally above the knees, back flat and stirrups a couple of holes or so shorter, depending on comfort and balance. The faster a horse goes the more slightly forward goes his natural balance, his centre of mass (gravity or balance) being situated inside his thorax/chest, directly below his spine two-thirds of the way down, about a hand's breadth behind the back of the withers.

The weight a horse feels on his back is mainly via the stirrup bars in any position in which the rider's seat is out of the saddle or just brushing it, so a well-fitting saddle should be able to be positioned so that the stirrup bars sit, one on each side of the back, of course, above or *just* in front of the horse's centre of mass for efficient balance, without interfering with the movement of the tops of the shoulder blades or pulling the girth too far forward so that it digs in behind the elbow. Both of these faults considerably hamper the horse's gait because they are very uncomfortable and can cause significant bruising and bad associations with being ridden. Most of the saddles I come across are placed too far forward.

The gallop proper definitely calls for shorter stirrups and a forward-carried torso, of course. It is now a regular four-beat gait because the second-beat step of the canter, for example, right hind and left fore when in right canter, and vice versa for left canter, is split, the footfall sequence with the right fore leading being left hind, right hind, left fore, right fore, in a regular four-beat rhythm with, of course, a moment of suspension.

Fast gaits engender excitement in most riders—and horses! I have a love–hate relationship with horse racing because it is a very stressful life for the horses, but I do watch it fairly often, and it disappoints me so much when I see how some jockeys ride a driving finish. So many fling their arms and hands all over the place as though this will make the horse go faster (which it won't), whipping them with cushioned whips which, I promise you, *do* hurt (and despite Australian studies having shown that whipping can actually slow horses down, I believe), and some jockeys banging their bottoms repeatedly down on the most sensitive part of the horse's back—just behind the saddle—riders with long thigh bones being the most prone to doing this. I do think that the 'driving finish' apparently taught at racing schools and elsewhere should be completely rethought with a view to producing a more effective and horse-friendly technique from the jockeys and improved welfare for the horses.

I was fortunate enough to be able to follow the horses in training with one of Britain's best jumping trainers, the late Gordon W. Richards, for a year, for the purpose of writing about them in my magazine, *EQUI*. The year I was visiting, he was training flat racers in the summer and jumpers in the winter, the only year he did that, and was always telling his jockeys and work riders to drop their stirrups a hole or two because 'legs were invented before whips'. Quite right. Now, though, stirrups are even shorter than they were then.

10   REGARD FOR SELF-CARRIAGE

Just as Andrew McLean has said that the correct use of operant conditioning is the heartland of training, he also has said that self-carriage is the vital element in training. The mores of both classical riding and ES emphasise that self-carriage is the ultimate goal in a correctly trained horse. Modern conventional riders, competitive and otherwise, may well have heard of it, but it seems to me that they are not even taught about it, let alone encouraged to aim for it. There is certainly no sign of *real* self-carriage, the kind I was brought up with, in what I see of the current competitive milieu. So—what does the expression 'self-carriage' mean?

Self-carriage means that the horse independently maintains his way of going in appropriate balance for his gait in everything he does under a rider or, in fact, during in-hand work. This involves maintaining, on a light contact and via cues/aids, his gait, speed/tempo, length of stride, line (more specific than 'direction'), head and neck carriage and general bodily posture—*rider permitting*. It is a real responsibility to ride a horse trained, or being trained, to go in self-carriage.

Just as the ultimate goal in classical riding and ES for a quality riding horse is self-carriage on the weight of the rein, the ultimate goal for an accomplished rider is an independent, balanced seat. An independent and balanced seat is secure and effective. The rider's balance accords with that of the horse, making his work so much easier for him, but the rider uses very little physical effort to stay 'in the plate'. Such a seat does not rely on the reins to stay on, so the horse's mouth is safer. It also does not rely on the rider propping with the hands on the neck of the horse when jumping, so the horse's head and neck are free, literally, to do their job and balance him over the fence. An independent seat also means that the rider is able to maintain it without one or both stirrups, if necessary, and it enables the rider to give cues/aids to best effect.

Except when using weight aids (*see* later in this chapter) or in any position in which the seat is out of the saddle, the rider's weight should be borne equally on the seat-bones, the insides of the thighs and the stirrups. To ride a highly schooled horse, the rider *has* to have excellent control of his or her balance in the saddle and of his or her arms, hands, legs, bodyweight, seat, position and use of those tools by being able to relax, brace and control the body as required at will: they must mentally know what they are doing and be physically able to do it. This is a goal well worth striving for because there is no feeling in the world like it. When you get it all together, you are as near to being a centaur as you will ever get.

In Chapter 1, I described Nuno Oliveira demonstrating weight-of-the-rein contact with a microphone lead. As you might guess, there is a lot more to this than simply letting the rein go and allowing the horse to work as he wants. The rein(s) can have a very shallow loop to them or a significant one but can easily feel a slight message from the rider via a little shake, vibration or change of the rider's hand position, up or down, out or in, or a turn of the wrist, with no other pressure in the mouth, and will respond.

Classical conditioning has been brought to its ultimate in this kind of riding: equally subtle aids are given via the legs and seat, the latter being doubted by some ES coaches, I believe, but nevertheless used extensively and effectively by good classical riders, even in the Iberian saddles which were the forerunners of the Western saddle. Even in the sometimes bulky modern dressage and flatwork saddles, the horse can feel the weight adjustment of the rider in weight aids and respond accordingly. It would help, though, if someone would design a saddle that would weigh less and be 'slimmed down' while still protecting the horse and allowing a comfortable and correct seat for the rider.

To bring a horse to the level of genuine self-carriage is a matter of strength and balance, in that order, because to bring his natural balance back a bit so that his hindquarters can take more of his weight (plus, don't forget, the weight of the rider and saddle), he needs to be strengthened or he could become injured. We work him by means of the systematic and regular performance of exercises that will gradually strengthen him and accustom him to carrying weight (no more than 20 per cent of his own weight). A programme of manège work alternated with, ideally, riding out, on five or six days a week is good. Exercises are used that intrinsically will lighten his forehand and get his weight back without coercion as his body gradually builds up strength (*see* Chapters 8 and 9).

I use the term 'self-balance' to indicate a horse not yet in what classicists regard as true self-carriage, as described. The point is that, from the earliest days of training, the horse must learn to carry himself and his (well-balanced) rider. This is no great ask because horses do not like being manipulated or falling down. A skilled, lightweight rider able to keep a light, steady, in-touch but not wishy-washy contact is ideal and can work alone or with a trainer on the ground. In the long run, it is quicker, and safer from an injury and behavioural viewpoint, to train a horse properly from the start. Such a horse will stay sounder for longer because he will be trained so that he is always working within his body's natural functioning rather than being forced into outlines and ways of going which necessitate using the wrong muscles for what he is being asked to do, and on a contact that actually stops him balancing himself.

THANK GOODNESS FOR THAT!

I have taught horses many times who have been former Prix St George and Grand Prix competitors, and none of them has been able to go in any semblance even of self-balance, never mind self-carriage, until I had been on board for a good 20 minutes, having first loosened their nosebands and lowered their bits to pain-free levels, and sometimes moved their saddles back a couple of inches so they could move their forelegs without discomfort. Gradually, within one lesson, they have adapted to the point at which their riders could get on and experience, for the first time, a horse moving freely, correctly, calmly, showing beautiful gaits and responding to aids quickly and lightly—no struggle, no hassle, just relief, surprise and enjoyment.

As explained, modern riding usually involves relentless pressures on a horse's sides with legs and sometimes spurs, and in his mouth from the bit, exacerbated by a tight noseband, which most probably cause pain, possible mouth injuries as described earlier, and make aids very difficult to discriminate.

It is no wonder not a few riders constantly nag with their legs or, worse, spurs to keep their horses going, haul on their mouths to stop and turn them, 'bit them up' with stronger and stronger bits and, at least at home, may use the whip to punish them when the unfortunate horses cannot work out what to do to avoid the discomfort or pain. It is also no wonder that so many horses react self-defensively against this type of riding because they just don't know what to do to stop it.

All this can be avoided by training using the principles of ES and real classical riding.

IN A NUTSHELL

Equitation Science is the science involved in horse training, riding and management. It is not a method or system in itself. Its ten First Principles have been discussed and explained here with my own comments. They are fundamental to the process of training and caring for horses and represent a huge step forward in doing so in a way that is comprehensible to horses and completely appropriate to the type of animal they are.

We have explored how horses really think and learn, which is not as we do. ES involves learning theory, ethology and cognition, biomechanics, physiology and sports science. It is a set of principles that we can follow to achieve best practice in our dealings with the horse and which help to ensure equine welfare. Lightness, self-carriage and the ultimate well-being of the horse are common to both ES and true classical riding.

The point is stressed that the advanced state of self-carriage on the weight of the rein has to be acquired by the horse naturally through proper work rather than being enforced by coercive training.

7 The Right Start for Raw Recruits and Old Hands

EQUINE LEARNING THEORY

Very many of the problems experienced by horses, handlers and riders are due to confusion on the part of the horse, and to lack of understanding on the part of the handler/rider/trainer of equine learning theory—how horses think and learn as opposed to how we have assumed in the past that they do these things. It is a scientific development, but many of its principles are long-established parts of classicism, too.

Equine learning theory presents principles and techniques that make it as certain as possible that horses are taught in such a way that they find it easy to learn because those techniques accord with how their minds work, which is differently from ours in some significant ways. Therefore, their lessons make sense to them and possible confusion is avoided if we apply the techniques and principles correctly. This achieves the best results and gives horses some control over how they react to what they experience, in this case the stimuli/aids/cues/signals they experience from handlers and riders, but it also places considerable responsibility on us—specifically, to get our consistency and timing of the aids right.

DOMINATION

Today a concept persists that causes not only confusion to horses but also distress, and that is the human tendency to feel dominant to other creatures and to need to be The Boss. This concept does not lie easily with horses, who do not live by such a convention in their natural lives either in feral conditions or a domestic paddock. There may be a 'dominant' horse for various resources, such as grazing, water, shelter and certain company, but horses do not understand a hierarchical society with one overall leader, such as we and some other animals, including horses' main predators, have: it seems they understand humans as part of their society but not as their bosses. They 'manipulate' each other in normal social engagements and horse play which, as we all know, can be pretty rough. They obviously understand their own often extremely subtle and sometimes violent body language, ranging from a look and an ear position to forceful kicks and bites.

Of course, we do assume a level of dominance in that we exploit horses for our own pleasures or business requirements, and if they did not do as we asked, they would be useless and maybe dangerous to us. Fortunately, by means of correct, virtually stress-free training that they understand because it accords with how they think and learn, we can create a partner, companion and, eventually, a friend who seems quite ready to do as we ask, as the initial training gradually turns into a delicate conversation—rather like putting our offspring through a good school. The aids, cues

DOI: 10.1201/9781003121190-7

or signals—whichever term you prefer—at first create mild to moderate irritation to stimulate the horse to perform the action we are asking for, and which the horse performs initially simply to get away from the slight but continuing pressure or irritation. Eventually, as the horse gets used to the cues and has learnt, by means of negative reinforcement (*see* later in this chapter), what to do to stop them, extremely light versions of them (such as bit contact or leg pressure) produce instant, correct responses almost by second nature or habit: the horse becomes 'classically conditioned' (gradually accustomed) to responding in a particular way to a particular aid or cue.

In most forms of equestrianism, horses are accustomed to humans and trained, little by little, from very early days, although it is advisable, to allow mare and foal to bond, to leave them alone as much as possible for 24 hours after birth. As in many other species, the temperament and character of the dam is a strong influence on her foal, and a calm, confident mare will promote similar characteristics in her foal, and possibly vice versa.

NEGATIVE REINFORCEMENT

Because negative reinforcement is a crucial component of good training that horses can understand, I'll give a layman's explanation of it before talking about procedures.

As well as leading, in any method the foal or youngster should be accustomed very early on to being touched all over, feet being picked up and cleaned out and so on, and this can be done with great advantage by using negative reinforcement.

For example, if you plan to start by touching his shoulder, put on his foal slip, have the foal near his dam, then place your hand gently but clearly on his shoulder. If he tries to move away from it, keep it on his shoulder till he is standing still, then remove it *at once* (his reward for standing still) and *at the same moment* say 'good boy' in a soothing tone. This reinforces/strengthens the likelihood of your being able to touch his shoulder again in the future and of the foal standing still. Rubbing the withers or upper shoulder, as described earlier, can also be done.

There is a major difference between ES-based negative reinforcement and conventional and also traditional methods in training a foal, for instance, to tolerate being handled. As an example, in ES the handler might place his or her hand on the foal's shoulder to accustom it to touch. If the foal permits this, the hand is removed and the foal rewarded, perhaps by a rub on the withers. If he does not permit it, the handler keeps the hand in place till the foal stands still, then removes it and rewards the foal.

In conventional and traditional training methods, the handler will keep on handling the foal to get him used to the contact, even after he has stood still, all the while praising him but not giving him any reward he understands on a practical level.

If the foal manages to remove himself from your touch, or if you remove your hand while he is still objecting, maybe because you don't want to upset him, you have rewarded his objecting behaviour and he will have learnt that he can escape your attentions, having reinforced his own action. All is not lost because future training can overcome this initial mishap, even though it is a backward step. Mature horses, for instance, can often be rehabilitated behaviourally to a large extent by this same method.

Negative reinforcement is used in a mathematical sense—'negative' means subtracting/taking away the contact or pressure and 'reinforcement' means strengthening the likelihood of the foal standing still in future, in the context we are considering as an example. The contact may be a bit worrying and irritating, so when it is removed, when the foal stands still, this acts as a reward and a confirmation that the thing to do to get rid of the contact is stand still. This gives the foal choice in his life, 'agency', and the knowledge of what to do to get rid of the unwanted contact. When the handler has said 'good boy', he or she can follow up immediately with rubbing or scratching the foal on the side of his withers or upper shoulder, which is the main spot horses use to mutual groom each other, as this will be pleasurable to the foal and is another good association.

Another important point to adhere to is to keep all aids/signals/cues as identical as possible so that the horse himself can rely on the trainer to be consistent. If negative reinforcement is well used, and it is used in all aspects of handling, groundwork and riding, and the words 'good boy', or whatever you prefer, are said identically each time, the horse will appear, in my experience, to be glad to do what you ask and take an interest in his activities with you.

POSITIVE REINFORCEMENT

This means the addition of something enjoyable such as a treat or titbit, or scratching or rubbing the withers/shoulder, as described. 'Good boy' can also come to be regarded as positive reinforcement. It can be given to reward a correct response to a cue and so increase the likelihood of its happening again. As ever, the timing is crucial (within a second) so that the horse understands that it is linked with his action.

Negative and positive reinforcement are vital aspects of all ES training and learning theory and are explained in full detail in all the ES books listed in 'Recommended Reading'.

TRAINING/SHAPING SCALES

This seems a good point at which to discuss the issue of training or shaping scales as guides to help us through the process and the responsibility of training a green horse or retraining an experienced one 'having difficulties with human interventions', as a colleague of mine put it. 'Shaping' means, in lay language, a gradual, logical progression, one stage at a time, towards a goal, such as in backing a horse or achieving light bit aids.

The FEI scale of training will be very familiar to conventional and classical riders and trainers. It was originally devised in Germany and later adopted by the Fédération Equestre International or International Equestrian Federation (FEI), and it is very widely used internationally, ostensibly as a progressive scale for training horses competing under their rules, in six essential qualities. It reads:

Rhythm > Looseness > Contact > Impulsion > Straightness > Collection

I recall a slightly different version of the present format years ago but, because it has been for decades the official FEI scale, it seems that everyone accepts it as the gold standard without, apparently, ever querying it when, in fact, it is not even actually a *scale* of what to teach and aim for. To be a scale, it needs to progress from one logical step to the next, slightly more advanced lesson to build on a horse's training in a rational way. The FEI offering is really a list of objectives, fair enough on its own but not a scale.

It starts off well enough with Rhythm followed by Looseness, which is good because neither of these is obtainable without the primary requirement of relaxation, but then it jumps to Contact and acceptance of the bit. I think it will be hard to achieve an even contact approaching lightness as the norm if the horse himself is not what we call 'straight', with his hind feet following exactly the tracks of his fore feet, and probably to a degree unbalanced, yet Straightness is not even next. Impulsion follows Contact and is itself then followed by Straightness.

Many years ago, I queried this with a well-qualified instructor giving me a lesson, who simply quipped: 'Well, I think the FEI knows better than you do', with no offer to discuss the matter. As mentioned already, the last thing we want is a horse going with impulsion if he is crooked, that is, not straight and maybe even with a tendency to wander off line, developing the wrong muscles and other tissues which can also be physically changed by a horse's movement (and that goes for us, too). After Straightness in the FEI scale/list comes its final quality, Collection, without any mention of important, necessary, intervening qualities such as, after Impulsion (which should be after Straightness), engagement, 'throughness' and the beginnings of 'lift' in the forehand.

The Equitation Science shaping scale, and it *is* a scale, reads:

Basic attempt > Lightness > Speed control > Line control > Contact > Stimulus [aid] control

- *Basic attempt* means that the horse has a try at the right response, even just one step.
- *Lightness* means that the horse gives an immediate, correct response (such as one stride, four footfalls) to a light aid.
- *Speed control* means that the horse himself maintains his rhythm and tempo (speed) in changes of gait requested by the trainer, and keeps going without being asked until cued otherwise (known as 'persistence').
- *Line control* means that the horse continues along the line chosen and indicated by the trainer *and is straight*.
- *Contact* means that the horse himself maintains connection with the passive, in-touch bit/rein contact but also the legs and seat of the rider; it also means that he maintains his head/neck and body posture, all independently of the rider.
- *Stimulus control* means that the horse is under the stimulus (aid/cue/signal) control of the rider or handler, responding every time, and only when asked, with all the preceding qualities and in all locations and circumstances.

After stimulus control, collection will present *itself* though correct work (bending and transitions, *see* Chapter 8) accompanied by the natural, *incidental* development of the sought-after 'rounded' outline, so much prized and the real proof of the training pudding. That is the rider's confirmation that the training has been correct—but it has to come as a by-product from the correctly working horse, not requisitioned by the rider.

That is the ES training or shaping scale and I commend it to my readers. But for now, we need to descend from the heady heights of collection and get back to basics.

OVERSHADOWING

One of the main problems with modern riding is that horses are usually subjected to two or more stimuli or aids at a time. When these are both asking for the same thing, such as a rein aid and a leg aid both asking the horse to move left, say in leg-yield, he can understand this and move left evenly. If, though, the rein aid is stronger than the leg aid, it overshadows the leg aid and the horse may move his forehand left more than his torso and hindquarters, losing at least some of the value of the exercise as he toddles off more or less just on a diagonal line.

Major confusion occurs when the aids are asking different things such as, commonly, a leg aid asking the horse to go forward and a bit aid telling him to slow down or stop (as described earlier). Horses' brains cannot work this out but if, for instance, the leg aid is stronger than the bit aid, it will overshadow the latter and the horse will respond to the stronger aid, going forward as required, if somewhat confused. If the bit aid is the stronger, he will restrict his gait and go with a strained, pulled-in head carriage.

Overshadowing is a double-edged sword because it also can be used for useful purposes such as helping a horse to overcome distractions in his environment. A common example used is in habituating a horse to clippers. If he is worried about clippers, his fear can be overshadowed by stepping him back and forward, back and forward, while an assistant slowly brings the clippers (switched off) closer. Your aids for forward and back must be light (*see* later in this chapter) before progressing and bringing the clippers closer. Then when he will let your assistant touch him all over with the clippers and your aids/signals remain light, you can start the same routine again with the clippers switched on. At no point must the horse be forced in any way to tolerate the clippers.

On any occasion when your horse is distracted, not paying attention or becoming fractious, think of something with which to overshadow the cause: stepping forward and back is usually very effective, ideally with the head lowered somewhat, if possible.

FIRST STEPS

From very early days, a foal can be encouraged to go where people want it to go much more easily if the dam is very close. A foal slip/light headcollar can be put on and the foal guided by two people, using stable rubbers or towels looped around its hindquarters and chest, and a lead rope attached to the slip.

It is not too soon to accustom the foal to vocal cues as well: the words 'walk on' can be said as the foal takes his first step so he learns to associate the words with starting to walk, while his forehand is steadied by a helper. The helper, at the same time, gives gentle forward pressure on the slip via the lead rope, which the foal will feel mainly on his poll. At the same moment, the mare, who should already know the command, is led forward by a second helper and the foal thus initially learns the command which is made easier because he will associate the sound with both he and his dam walking forward. After only a few days, he will walk forward independently in response to gentle forward pressure on the lead rope and can be led calmly around his dam, then a little further away from her, and so it progresses. An excellent start.

The practice of allowing foals to run free around their dams but basically following them is common but invites problems later in getting the foal to respond well to handling and to respond accurately to vocal cues. Of course, this happens in the field, but when humans are present, it is not a good idea.

BACKING

Backing a horse, that is, getting him used to having a person on his back, wearing a saddle and girth, and being led around with the rider on board, is something conventional trainers have their own myriad ideas about. Usually, the horse is accustomed to a saddle pad, then the saddle is gradually introduced in the same way—and the girth, which usually causes the most problems (see later). If the horse has been lunged and/ or long-reined, he will be used to wearing a roller for attachment of the hopefully comfortably loose side-reins, and the girth will cause few if any difficulties. Another preparation method is to use an elasticated surcingle/body girth, applied very gradually around the thorax where the saddle and girth will go, in several steps before it is fastened just tightly enough to keep it in place, then the horse led round wearing it.

The trainer will benefit from a helper not only for safety reasons in case of an accident but also because it makes the learning of ridden aids easier. If, during the weeks preceding backing, the groom/handler has made a practice of firmly stroking and brushing the horse in the saddle and girth area, putting some weight on the horse's back and jumping up and down next to him, with his or her hands on the back taking some weight, this will have gone part way to helping with the backing process.

The ES approach to backing and accustoming a horse to the bridle, saddle and girth is similar to the traditional way but more specific, structured and reliable, in my view. It is described in detail in *Academic Horse Training* (see 'Recommended Reading'), but I'll give the main points here.

The horse will be used to wearing a headcollar, and the usual way of getting a horse to lower his head, and accept a bit, have been described earlier, so those three points together, effective in practice if done correctly, will make fitting a bridle and bit fairly easy. A very detailed method of head control, though, is given in *Academic Horse Training*.

Backing itself is done without the saddle and through several stages, from placing weight on the top of the horse's shoulders, making small jumps next to him and being held up with one leg by an assistant, on to lying across the horse's back, placing a knee over his back, then a leg, and finally sitting up. Each stage requires at least three repetitions during which the horse remains calm before moving on to the next stage.

The handler holds the horse's head and will step him forward and back, as in the in-hand training (see later), which helps to accustom the horse to the process.

The procedure continues with the handler controlling the horse using the familiar aids from his in-hand work, leading around and making use of stepping back and maybe head down, which always seem to have a good calming, focusing and over-shadowing effect on the horse. The rider gradually takes control as the horse learns the ridden aids in association with the in-hand ones.

The care taken to ensure calmness and correct responses at each stage and the insistence on good repetitions to confirm the work make the process much more reliable and less potentially fraught than backing and riding away can be in conventional riding, in which the horse is often not given the time to absorb the aids, and the aids are used illogically by the trainer due to lack of knowledge of learning theory. Negative reinforcement is not usually used because it also is not known about. In the effort to make a horse ready to sell and compete, work also may be rushed and forced by some 'producers', to the detriment of the horse's understanding, well-being and the safety of the horse's future riders, due to his defensive reactions to confusion. This may not be widespread but is common enough to be a problem, judging by the number of wary, unresponsive and defensive youngsters, and older horses, I have been called to help with.

BITTING

A green horse should not have side-reins or anything else fitted to a bit in his mouth. If it is decided to use comfortably and safely loose side-reins, at first they should be clipped to the side rings of the cavesson. Only when the horse is quite accomplished at lungeing, calm and cooperative should they be clipped to a bit, and a riding bit, not a 'breaking' bit—*and* only if the horse has been bitted properly.

Special mouthing bits with loose keys ('danglers') are meant to encourage the horse to play with the keys and get used to the feel of the bit, but they can also prompt him to get his tongue over the bit, so it is better to use an ordinary eggbutt snaffle, with a lozenge rather than a single joint, in some tasteless material. These mould best to the shape of a horse's mouth and the eggbutt joints prevent pinching of the corners of the horse's mouth.

When introducing the bit, it should be fastened to the bridle at one side only, and a finger inserted in the horse's mouth on the other side (in the corner of the lips, of course, where there are no teeth). This will open his mouth a little and the bit can be brought through into the mouth, across the tongue and fastened to the other side.

The old process of bitting is very stressful to horses and can cause a good deal of tension and so pain in the muscles of his neck and shoulders. In this procedure, the horse is fitted (and I say 'is' because sadly it still happens) with a bit and a roller, side-reins fastened to both and the horse left standing tied up in his stable for increasing lengths of time to get on with it. As well as being too stressful, it is also dangerous as the horse might begin to fight his situation, fall and be unable to get up again: appallingly bad horse management. Mentally/psychologically, such a distressing experience, fall or no fall, can affect a horse's whole attitude to equipment, training and people in general.

Horses being bitted should never be left alone, strapped up, tied down or tethered. A sensitive, competent trainer should put the bit in the horse's mouth (no side-reins) and stand with him for a very few minutes, calming and stroking the youngster and perhaps giving him a treat such as a hard mint, which most horses like, to make a pleasant association and create movement of the jaws as he gets used to the feeling of a bit in his mouth. In time, he can pick a little grass in it, too.

He is less likely to start trying to get rid of the bit and put his tongue over it if the wearing times are very short at first. Gradually, the time can be increased and the youngster led out around the premises from a cavesson fitted with the bridle. The trip will also take his mind off the bit. The bridle and bit are put on first, then the cavesson is put on top and the lower parts of the bridle cheekpieces undone and brought up over and outside the cavesson's noseband, then fastened again. This prevents the cavesson interfering with the bit and allows a less fixed position of the bit in the horse's mouth, which seems more comfortable for him. No riding reins are fitted, the horse being led around from a lead rope fitted to the nose ring of the cavesson so as not to pressurise the bit in the mouth of a green horse but offer good physical control to the handler. Alternatively, the rope can be clipped to the side ring on the cavesson, changing sides occasionally to maintain ambidexterity. In this way, the horse can be distracted from the bit in his mouth and get used to it gradually for short periods without distress, pain or obvious danger.

MOVING ON

Let us imagine a young horse who is well handled and ready to become accustomed to wearing riding tack and having a person on his back, and look at the methods used in classical riding, conventional practice and Equitation Science training.

NB: As this is not a manual on all the techniques used in training/schooling horses but about blending the principles and (some) techniques of classical riding and Equitation Science as a detailed introduction for readers to one or both, the principles and practices I have chosen to write about are those which are particularly good, particularly bad or which merit discussion.

First, the expression 'breaking in' is an unpleasant old phrase from a welfare impression and can be so in practice, but most people know that it means getting a young or green (untrained) horse used to the previously mentioned processes, and 'backing' just means getting him used to a saddle on his back and then a rider. 'Riding away' means getting him to walk and trot, and later canter, under the rider in response to his or her aids/cues/signals. The ES expression for this process—'foundation training'—is much better and more appropriate.

Good conventional and true classical horsemen and women will use the traditional approach, often with very good results. The pattern will be fairly familiar to most readers who, however, if having little knowledge of ES, may not have realised some of the potential disadvantages of the traditional method, or the advantages of the ES approach.

Figure 7.1 Anne Wilson on her lovely Spanish mare, Secret, having come to a beautiful, almost perfectly square halt—happy horse and rider, great balance and comfortable bridle. Secret came to ridden work later in life, having been a broodmare for some years, and has taken to her new life beautifully.

HANDLING AND GROUNDWORK

Handling in conventional and classical schooling will have taken place since foalhood and vocal cues or aids used to stimulate a gait, such as, for example, 'walk on' and 'trot', but we have naturally tended not to be precise about our use and timing of aids. They should not be used to keep the horse going, as is common, because this is confusing when he's already complying. Only repeat the vocal and physical aids/cues if he slows down significantly, or actually stops, and as soon as he does so.

We don't need to keep reassuring him, either, with 'good boy' that he's doing the right thing. If we said 'good boy' as soon as he complied, that's fine, but later on it's just excess baggage. It could just confuse him, and confusion is probably the most common feeling that we engender in horses in conventional and traditional equitation. Horses don't vocalise to each other all the time and neither should we. We should just stand quietly and let the horse continue until we signal him to do something else. This was also a training requirement of the classical masters of old.

The important point is the timing of the application and release (stopping) of our aids/cues/signals. Whatever school of thought we're following, we *must* cease the aid, whether it is physical or vocal, as soon as the horse complies, ideally within one second, for the horse mentally to be able to connect the aid and the act. This is another old classical requirement—remember Dési's salt remark—which modern research has confirmed as correct.

Research seems to indicate that horses cannot think forward or back. If we fail to give our reinforcement/reward as soon as the horse does what we are asking, in other words if we keep giving our leg aid, our bit aid or our vocal aid even though the horse has complied with it, he could conclude that whatever he did didn't work (as in getting rid of the aid). He is trying to relieve himself of the slight irritation or pressure of the aid, remember, and will try something else ('trial and error' learning). If we *then* stop giving our aid because we were too slow, we shall have reinforced/rewarded the second, unwanted movement that he tried, because the horse will connect the aid with that, which is not what we want.

So the point here to remember is that *the horse will connect the release of the pressure/aid with whatever he was doing immediately before we released.* We have to concentrate and stop the aid *immediately* for the horse to learn how to respond to it. That's our responsibility and it might be a new concept to conventional, modern riders, but not to true classical riders, and ES equestrians will do it automatically. Don't worry—it will become second nature to do this after a while because we become classically conditioned to it!

It is also important to lead foals and youngsters from both left and right sides and on the left and right sides of the dam because making them ambidextrous will make all aspects of working with them in future so much easier. As the foal progresses, having the leader between him and his dam also accustoms him to a bit of independence, yet his dam is still nearby.

TRAINING AIDS

Most people in the *conventional, modern sphere* use training aids, or side-reins attached from the side rings of a lungeing cavesson or even a bridle and bit to a saddle or roller, as an extra means of control and to place the horse in a 'rounded' outline. These tack arrangements, however, are often fitted too tightly or short to hold in the head and neck and compress the body so that the horse is unavoidably held in an artificially rounded outline, with a shortened neck and head-in carriage. The general belief seems to be that working in this posture will develop the muscles to adapt to it (in practice, the wrong muscle use can cause discomfort, pain and injury), but I find that most people do not understand why this posture is desirable (apart from appearance which some find attractive) or that it cannot be developed correctly by such coercion.

The rounded outline and head and neck carriage are the *result* of correct training (without tight gadgets or hard, unrelenting bit contact) as the horse uses his musculature naturally, and to best effect, to maintain the slight rearward shift in his balance/weight that is needed to lighten his forehand and make him light in hand and agile. To counteract this and adjust his own balance, the horse naturally pushes his neck and head out and down in the early stages and, as he becomes able to take a little more weight behind, the neck and head posture changes to being stretched up and forward in an arched shape with the head vertically flexed at the poll so that the front of the face is either on the vertical or just in front of it. It has nothing to do with showing spirit or being proud and looking good but everything to do with the natural need of the horse to keep his balance.

The fact that many people find the posture indicative of spirit and pride is incidental to sheer practicality, from the horse's point of view. *Very* few people seem to understand this, so competitors manipulate the shape or posture demanded to gain marks for it. I think that this should be seen as a welfare issue if, as I think likely, it puts the horse under inordinate stress and strain. An artificially acquired neck and head carriage is certainly not a mark of ethical, correct training.

Classical trainers may well use side-reins (but not usually other equipment) adjusted so that they only come into contact with the horse's mouth (if attached to a bit on a bridle) or nose (if to the side rings of a lungeing cavesson) if his head moves significantly out of a natural position for his gait—an understandable precaution with a green horse or one being retrained. Depending on the horse and his stage of training, it is very effective to lunge a horse with no gadgets or side-reins at all, and rewarding to watch his physique and posture develop as he uses his body naturally in correct work decided by the trainer (to be explained later).

LUNGEING

Apart from concentrating on training our youngster to be handled and to lead calmly and safely, not much progression will have taken place in classical, traditional and conventional, modern practice until the youngster is 3 years old, although some schools of thought begin earlier, such as in Thoroughbred flat-racing in which horses are 'broken' as yearlings so that they can be ready for the racecourse as 2-year-olds, which is regarded as too young and stressful by many other horsemen and women.

A common approach to actually starting a youngster and beginning his training as a riding horse is to teach him to work on the lungeing rein, teaching him commands and developing his gaits. For this, preferably he should be fitted with a stout, well-fitting lungeing cavesson, one that will neither pinch nor pressure him unduly and that will stay in place and not twist round from the pull of the lungeing rein on the nose ring. This can occur just from the rein's weight, pulling the side/cheek piece into the corner of the outside eye and distracting and discomfiting the horse.

In ES training, lungeing is not habitually used and not needed and neither are training aids.

The word 'lunge' comes from the French '*longer*' which means to work a horse on a long rein no less than 6m or 20ft in length. It is also usual for the trainer to have a long lungeing whip with lash and thong which, with practice to develop the skill, can be run out to touch the horse to keep him out on the circle, stimulate more energy or slow or halt him. It is used as a guide to the horse, never to cause him pain or distress.

Lungeing does not have to be done entirely on a circle. Be prepared to walk with the horse on straight lines and large ovals, as well, rather than being rooted to the spot. A helper can initially lead the horse on his outside from the cavesson or rein to keep him out on the circle, pushing his head gently but clearly along the line of any curves being used such as circles, ovals or shallow corners, as she walks level with his head or neck, but with the vocal, body and whip aids coming only from the trainer to keep things clear for the youngster. As he gets used to the motion of the whip and the direction in which it is pointing, he will learn to stay out himself and, hopefully, look where he is going, and the helper will not be needed.

He will work from the in-hand vocal aids he has already learnt, and it is crucial that he slows down or stops to whatever vocal aid the trainer has used previously.

This is particularly important because very many non-classical trainers lunge horses almost entirely on circles which are often much too small, and they encourage the horses to go much too fast in the mistaken belief that this is creating 'forwardness'.

- *First*, 'fast' is not the same as 'forward', which means that the horse is 'on the aids' or ready to move promptly in whatever way or direction the trainer requests, that is, as the aids/pressures indicate. Excessive speed just puts a horse out of balance and can really frighten him, triggering his survival instinct and his flight-or-fight response. When the flight-or-fight response is triggered in any context, because it is part of the horse's survival mechanism, it can be impossible to eradicate completely and the horse will associate it with the context—place, activity, person, tack, position of trainer etc.—indefinitely, so every effort must be made to avoid it.
- *Second*, if a horse falls out on the circle, usually looking outside it, say to his left if on a right circle, it is usually because the circle is too small for the speed at which he is travelling and he is out of balance, which is scary for him. Horses use their heads and necks, of course, as we use our arms, to balance themselves in movement, and that is what such a horse is trying to do. Lunge reins are the length they are for a very good reason—to give the horse ample room and a gentle enough curve to work *comfortably, calmly* and *correctly*, with the trainer holding the rein at almost its full length with, perhaps, one loop in the hand. It is injurious and dangerous to lunge horses (or work them in hand) on circles that are too small. The horse's expression and attitude of face and head will reveal whether he is comfortable and feels safe.

The slow/stop aid from the lungeing rein (usually a vibration sent down the rein by the trainer) feels nothing like the in-hand aid the horse has learnt already or, of course, the one he will experience under saddle, so it is even more important that the vocal aid is very familiar to him and identically spoken each time, and that he responds correctly. He will pick up on the different context (in-hand, lunge or ridden) in time. The position of the trainer level with his shoulder, or head if necessary, and holding out the whip in the forward hand pointing in front of his head to slow him down, is also important. Every effort must be made to nip excess speed in the bud.

Fast gaits nearly always excite horses, but putting them in a situation in which they cannot maintain their balance is very frightening and dangerous. In addition, the trainer, to encourage forward movement, may be standing probably level with the horse's hip in a 'driving' mode, and holding the whip out towards the horse's hindquarters, a scenario that can easily be perceived by the horse (of any age) as being chased. As a prey animal, this is perfectly understandable and must be avoided.

Keep everything slow and calm. Only dispense with your helper when the horse is reliably responding to your aids/signals of voice, whip and body position, and posture: do not be tempted to lunge on constant circles, small circles or too fast, either. Lungeing involves a lot of working on curved lines which is hard work for the horse, using relevant muscles for a longer period of time than in-hand or at liberty, so short spells on a curve should be used, interspersed by straight lines.

Lungeing may become a permanent feature of a horse's life. Many people use it when they haven't the time or inclination to ride, or to warm up a horse before work;

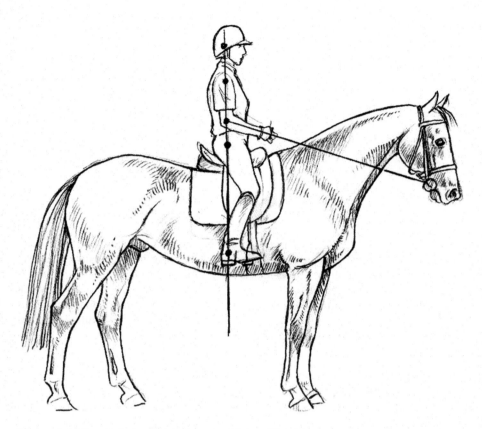

Figure 7.2 The default classical seat which will be familiar to many readers. The line passes through the crucial joints which, when aligned in practice, provides excellent balance, enabling accurate and easier weight distribution, making life easier for the horse, and aid/signal application. From the top of the head, the line passes through the ear, the shoulder, the elbow held back at the hip, the hip joint itself and the flexible ankle joint which absorbs concussion. Note that the line does not pass down the back of the heel, as is sometimes stipulated, because this can cause the lower leg to be held a little too far forward.

in such cases, then, it is most important that it is regarded as an important skill that can have a great effect, for better or worse, on a horse's attitude, not just a throw-away practice to get the itch out of a horse's heels, because so much can go wrong and permanently adversely affect the horse's mind and body.

LONG-REINING/LONG-LINING

Long-reining, or long-lining as it can also be called, is usually used after a horse is competent at lungeing and is less popular with some trainers because they have to move around a lot more with the horse. It helps to keep you fit, though! A horse's

Figure 7.3 How *not* to lunge. Lungeing can be seen as a 'chasing technique' and, so, inappropriate for a prey animal. The horse in this picture is clearly having problems! He is upset, being lunged on too small a circle, and is restricted by tight side-reins, so is well out of balance. His head and neck are turned out to the left as he tries to keep his balance and stay on his feet. He is also being lunged from the bit which is really the domain of experienced, knowledgeable and sensitive horse trainers.

performance skills can be brought to a much higher standard with long-reining, the trainer can move from side to side of the horse without altering the equipment, and get much closer to him. There are various ways of long-reining within conventional and classical equitation, of which I think the English one is the worst. There are the French and the Viennese, but the one I prefer, having learnt it from the late Sylvia Stanier, is the Danish.

The *English system*, however, is possibly the most widely used internationally. In this, the horse wears either a roller with side rings or a saddle with stirrups, which are down at normal riding length and fastened together under the horse's breastbone. The horse can wear a cavesson but usually a bridle with snaffle bit and no noseband or riding reins. The long reins are clipped to the bit rings and passed back through the stirrups to the trainer's hands. There are individually preferred variations on the equipment, but, in all cases, the common drawback is that the reins run rather low and their weight can encourage a horse to overbend and go on the forehand. Rein aids also can be hard for the horse to discriminate and it is hard to achieve a light, communicative contact. The horse may initially object to the reins running against his hind legs and needs to be correctly habituated to this, negative and positive reinforcement, as described, being excellent for this. A lungeing whip, or sometimes a driving whip, is used for guidance because the trainer moves at two or three metres (3yds 3ins) behind the horse. This way of long-reining is used mainly for novice horses.

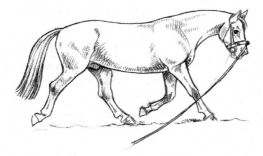

Figure 7.4 A good way to lunge. Here is a calm, relaxed and interested horse, trotting on nicely wearing only a lungeing cavesson (although most of us would also fit boots all round and over-reach boots in front). His head and neck are held naturally and, if trained to the command 'head down', could be asked beneficially to drop his head a little, which would naturally raise his back slightly and bring his hindquarters under, as he gains strength and experience. He is looking slightly around his circle and can develop correctly because he has *no* equipment restricting his posture and action. Bending work, which includes work on curves like this, is one of the elements of lightening a horse's forehand and very gradually strengthening his hindquarters and legs, provided he can work freely and correctly, like this.

In the *Danish system*, the horse wears a driving pad or roller with raised terrets (metal rings or loops on short posts), usually three down each side of the roller so that the height of the reins from the bit or cavesson to the trainer's hands can be varied—the higher the reins, the more advanced the horse, the highest being at riding height. The rider works closer to the horse and the terrets also take some of the weight of the reins off the horse's mouth. This position of the reins also enables the trainer to move easily from side to side of the horse. Horses can be trained from preliminary work to High School and achieve a very light contact which allows them to use their bodies naturally and develop strength and the quality of their gaits.

Lungeing and long-reining, and other 'loose work' techniques, need knowledge and common horse sense if they are to be carried out beneficially. The problem with all such techniques is that they can so easily be seen by the horse as being chased which, for a prey animal, is one of the worst things that can happen. None of these techniques should be regarded as 'easier' than riding. They can all be potentially very damaging both physically and mentally, and it's a good idea to get truly expert advice and study them yourself before getting involved.

There are two little illustrated books by Sylvia Stanier listed in 'Recommended Reading' on lungeing and long-reining which are now definitive works on the topics.

INITIAL GROUNDWORK IN EQUITATION SCIENCE FOUNDATION TRAINING

There are four main basic movements or aspects of handling and riding that are important in themselves and that form part of all the other movements we ask of horses such as lateral exercises, jumping and more advanced work. The basic movements are to go forward, to stop (including slowing down and going backward), to turn the forehand both ways and to turn the hindquarters both ways.

Part of foundation training according to ES involves what I have found to be a more or less fail-safe way of training horses to do these things in-hand before they are backed. The principles and techniques can then be transferred easily, quickly and logically to work under saddle, with both young, inexperienced horses and old-timers who have learnt some bad or, rather, defensive responses to our attempts at training and riding, through no fault of their own.

I have been asked to help with some horses whose behaviour has badly frightened their owners to the point where they dare not ride or, sometimes, even handle their horses, but it has all been down to that C-word—confusion—including the fear caused by the rough, tough methods applied to them by people called in to remedy their situation and which, judging by some owners' descriptions of their 'remedies', amounted to abuse. Once the horses I worked with had gone through this simple and rational ES way of learning the basics, all but one, whose owner did not have the confidence to adopt new ideas, have become calm, cooperative and apparently as safe as an animal like a horse can be.

An important point to make is that if a horse has been badly frightened by a particular stimulus, from a jet aeroplane to a savage dog, road works to a braying donkey (believe it or not), or to ill-treatment in a particular context, even after retraining he may revert to his defensive behaviour if that situation arises again and he is frightened enough. This does not always happen, a good example being that a horse involved in a trailer/loading accident or a road accident in transport may load and travel in future without problems.

So, we'll start with moving backward, then moving forward, then turning the forehand, turning the hindquarters and finally standing still without being held, described as 'park' in ES terminology.

How to Do It

Equipment: You need to wear your hard hat/helmet, strong shoes or boots and gloves, as usual, and a body protector if you wish. You will also need a schooling whip. It is essential that we realise and accept that the whip is an aid to communication and understanding, so that we can touch the horse on parts you cannot otherwise reach: it is most certainly not for punishing or causing him fear or pain. It is used to create minor irritation to get him to move away from it and so relieve himself of it.

If there is any doubt about this, accustom him to the whip by using negative reinforcement, stroking him gradually, gently (but not tickling him) all over his body (starting on the shoulders) and legs, and not taking the whip off him till he stands still, then removing it immediately, rewarding as previously described.

Your horse just needs to wear a close-fitting but not tight, ordinary headcollar (not a controller-type headcollar or one exerting additional pressures) or, if he has been bitted, a comfortable snaffle bridle, my general preference out of the many hundreds of different designs now available being an eggbutt snaffle with shaped cannons and a lozenge in the middle connecting them, which will mould over the shape of the horse's tongue and lower jaw/mandible. I prefer taste-less bits because you cannot be sure your horse will like whatever flavour the manufacturer has added, if any, and to give the horse a hard, non-sticky mint, if necessary, *before and after* inserting the bit, to relax his jaws and create a good association with the bit. He will also be thinking more about the mint than the bit but, if you make the one-before-one-after regime regular, he will not expect more or start mugging you.

Environment: Distractions should be kept to a minimum so try to have a quiet environment with no noisy children, radios or clattering gear on the yard. A large, empty loose box would be good, with no bedding so he can move his feet easily. An outdoor pen is also good or you can use an outdoor or indoor school but not, I sug-gest, if other people are also using it. Horses do tend to learn better in the presence of others, but they should be standing quietly by. The coming process will be quite new to the horse, and maybe to you, too, so a quiet and calm atmosphere is important.

Figure 7.5 One of Sylvia Stanier's pupils demonstrates Sylvia's preferred Danish method of long-reining.

Procedure: First of all, you are going to teach your horse to go backwards, reliably, to a light feel via the rope on his nasal planum (the flat bone down the front of his face) or via the reins from the bit in his mouth. Place him with his right side next to the wall or fence and plenty of room behind him to step backwards. Stand on his left side by his head, facing his tail and almost arm's length from his head/top of his neck. Double up the lead rope so it doesn't hang down, holding it in your left hand about 15cms/6 ins from his mouth behind his lower jaw. If he is wearing a bridle and bit, bring the reins over his head and put your left thumb in the loop at the buckle end to keep them up, then hold the reins in your left hand, again about 15cms/6ins from his mouth behind his lower jaw; I like to separate them with my forefinger. Have your schooling whip in your right hand.

To Train Going Backward

Now you are going to ask him to take a step backward by giving the aid on his face or in his mouth. *Do not give a vocal aid or say anything; just stand still*. The aim is to get him to work out and correctly respond to the pressure aid himself, not to

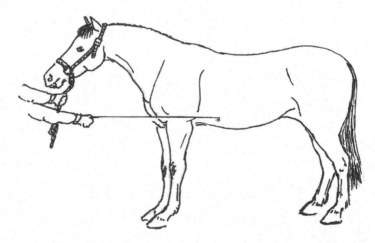

Figure 7.6a Diagram shows where on the horse's side to tap with the whip to train moving forward.

accompany or follow your movements, so don't walk in the direction you want him to back up. Be happy with one step, moving a single foot, at first. When you reach the point where you have to walk to stay with him, make sure he moves first, then you follow his movements.

- Direct the rope/reins *straight* backwards towards the underside of his neck, horizontal to the ground, and exert a light, clear pressure on it.

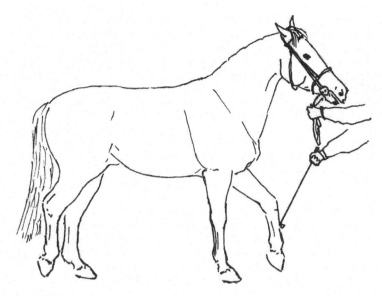

Figure 7.6b Diagram shows where on the horse's cannon to tap with the whip to train moving backward.

- If he does not step back within a couple of seconds or at least make a 'basic attempt' by lifting a hoof and putting it down a little way back, *don't* release (stop) the pressure but increase it a little and then, if necessary, vibrate the aid lightly and quickly, keeping the rope straight and horizontal to the ground.
- *Do not stop if he doesn't comply.* Keep vibrating (not jabbing) and also *tap* quickly, about two taps a second, with your schooling whip on the front of his left cannon continuously, speeding up the tapping if he doesn't respond.
- It is vital you keep up these aids until he makes a basic attempt to get away from the irritation of the aids by picking up his left fore hoof and putting it down again a little way back or, of course, taking a full step backwards. It might take a minute or so this first time, but continue without a break. If you stop even momentarily, you will have rewarded him for *not* stepping back.
- When he makes a basic attempt, even if the hoof only lands a short distance back, stop both aids *immediately* so that he knows he's 'found the switch' that will stop that irritation, say nothing, but quickly scratch or rub the side of his withers or upper shoulder as a pleasant reward. Some trainers then like to add the words 'good boy' straight after you start scratching.
- Now give yourselves a brain-break for up to five seconds, then repeat. It will happen quicker this time. Do five to seven repetitions in exactly the same way, then stand or walk him around, head low, for about 1½ minutes before doing a second, shorter set of three to five repetitions. Repeat the rest period, then do a third set of repetitions of one to three repetitions.

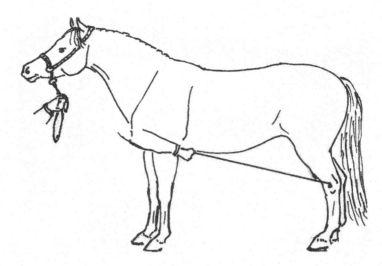

Figure 7.6c Diagram shows where to tap on the left hock to train moving the quarters
         to the right.

You'll find that your horse will get better and better at this technique. You can soon
stop using the whip-taps and, ultimately, he will go backwards from a single light
pressure aid from the rope or rein. Of course, your horse has two sides! So you'll
need to work the same procedure on his other side with everything reversed but with
the same timing. I find it best to work on one side in the morning and the other later in
the day, or the next day. It does vary and is not inflexible: I have taught this from both
sides in one day. I would say, also, not to ask him for more than two full back strides,
eight footfalls, and only once he is confident with that as the days go on. Never try
to rush or force the horse, or run him backwards. Training anything is always more
successful done in short sessions, keeping everything calm, clear and stress-free.

Of course, if he takes two strides back, you will need to walk forward with him,
but he must move first. If you are training in a box or small pen, you will also need
to train him to walk forward in the same session or you'll have nowhere to go!
Although this is not ideal, if you remain very clear and calm, with a minute or two's
break between going backwards and forwards while the horse is learning, it is perfectly feasible and I have done it many a time.

To Train Going Forward

- Stand as before although you could be a little further back just in front
  and to the side of the horse's head. You are going to ask for a single step
  forward. Exert the same level of pressure on your rope or reins *straight
  forward under his chin*. He will raise his head a little and feel most of the
  pressure on his poll, although, from the bit he will feel some on his lower
  jaw. Remember, no vocal aids/cues. Let him learn one thing at a time.

- As before, start with a light pressure, increase slightly if needed, and if necessary vibrate the aid—all within three seconds. On the fourth second, I find, you will need to tap his side with the whip where a rider's leg would go and, again as before, keep up the pressure, vibration and tapping till he takes a step forward. Then stop them *immediately*.
- Although these pressures are different in one way from the backing ones, I find that horses seem quite able to transfer them to going forward, and it usually happens quicker than backing. Don't step back yourself until later in this training (because he will be walking further forward). The object is to get him to move to an aid on his side, not to your leg movements on the ground.
- Release your aids immediately on the correct response, and reward by rubbing his withers again.

The repetition schedule is the same as for backing and, obviously, you will need to carry it all out for both sides. Ultimately, use the very lightest rein/rope pull, but continue the tapping till he moves, to simulate your upcoming leg aid when riding.

To Train Moving the Forehand to the Right

- Walk forward as normal, leading your horse on his left side.
- If he is wearing a headcollar, you cannot give a direct turn aid to the right but can pressure the noseband area on the left to push the head over by giving your aid with the rope to the right under his lower jaw.
- Also, it helps to tap him on the shoulder behind your back with the whip, or get a helper to do this perhaps more accurately, ideally just above his elbow, coordinating your timing.

Repeat, rest and reward as before, although the horse will probably pick this up very quickly because you will have been leading him around previously. This method, though, is more precise. Of course, then train to the left.

To Train Moving the Hindquarters to the Right

- Stand your horse in the open, standing to the left of his neck and holding his head with rope or reins.
- Tap with your whip on the outside of his left hock to encourage him to move that leg under his body to the right.
- As before, keep tapping till he moves the leg: it can help if you turn his head slightly to the left.
- Stop as soon as he makes one step, reward and rest.
- In time, you can progressively ask for more steps, and gradually move the whip-tap area up on to the side of his hindquarters.
- Never tap more forwards in the flank area as this is too sensitive and can trigger running backwards and kicking out.

Train to both sides and, if you wish, introduce the vocal cue 'over' as the horse starts to move the leg in the correct direction.

That Australian doyen of horse training, the late Tom Roberts, whose books are treasures to add to any equestrian library, said that a horse must *stand* still, not be *held* still. How very true. How many horses and ponies do you know who don't do this and whose owners don't try to do anything about it? Yet it is such a boon to have a horse who calmly stands foot perfect as you work around him or, crucially, mount, ideally from a mounting block or convenient wall or rock high enough for you to just step over on to your horse rather than stressing his back and saddle by using a stirrup, especially from the ground. The husband of a client of mine had such a block, with steps, made for his wife when she started having hip trouble: she said she wished she'd thought of it years earlier and how much kinder it was for the horses.

Having a horse who will *stand still* or *'park'* on command is the same as having a dog who will stay. (If anyone has ever trained a cat to sit, stay and come, please let me know!) Really, it is a basic requirement for reasons of safety for horse and human, let alone convenience, but also it really calms horses down to know they are not required to do anything. It quietens fidgety horses and seems to increase a horse's self-confidence, I find. Bear in mind, though, that if anything sufficiently frightening occurs, the horse will probably move! So let's look at the ES way of getting a horse to stand or, in ES parlance, 'park'.

To Train Standing Still or 'Park'

To learn this, your horse must have been trained to stop and go forwards and backwards, so being familiar with the rope or rein aid and, crucially, the whip-tap. The objective is to have the horse standing still until he is asked to move by means of a rein or rope aid, not in response to a voice aid or by following your leg movements. Many people, including me, do like their horses to respond readily to their voice aids and this is understandable, but the rationale is, I believe, that the horse could move off if he hears someone else saying a familiar command.

- Lead your horse around your arena in walk, then halt him with the usual light rope or rein aid. He will not need a whip aid to halt at this stage but carry the whip as you will need it soon.
- Holding the rope, or reins over his head, take one step backwards away from him, being sure not to exert any pressure on the rope. If he steps forward to start following you, immediately tap him back on the leg that is most forward, not just to halt but step back to where he was, hopefully only one step back.
- As soon as he is back in place and standing still, rub his withers.
- Walk back again, being ready to tap him back if he walks forward. If he does, tap him back. He will be catching on by now.
- Try again and walk back to the extent of the rope or reins, if you can, *and without an inkling of pressure on them*, being ready to tap him back, and stand him still, as before, rubbing or scratching his withers.
- Next try walking from side to side in front of him, always being ready to correct him, and ultimately walk around him on both sides, as far as the rope/

reins will let you, maybe stroking him as you go because you might use this technique for grooming or tacking up.
- Eventually, you will be able to run all round him, both ways, and although he might watch you, he will not move a foot. You need to consolidate your training in other safe areas.

There is, it's true, a great deal to be said for teaching a horse to stand still on hearing a vocal command. It can be invaluable if a horse gets loose somewhere he shouldn't, for instance, such as on a road or showground. If you want to train this, give your chosen command as soon as he stands still when you have tapped his legs, so he can associate the sound with standing, then every time you want him to stand use it and tap him back into place if he moves. Pretty soon, he will become classically conditioned (used to) standing still till you cue him to move. The safety issue also comes in because you have trained him to only move off in hand, after standing, to a forward-pressuring rein or rope cue—no vocal aid such as 'walk on'. If you train your horse this way, you need to tell anyone else who handles him in your absence, otherwise your horse could become confused—and it may not be easy to get through to them to do this if their habit is chattering to horses.

Equitation Science is evolving all the time and you might come across slightly different versions of the previously mentioned ways in the future, in books, articles and lessons if you use an ES trainer, and on DVDs, online and elsewhere. I recommend not changing just because of something you have heard or seen on social media, but to check on it first with an appropriate trainer or the ISES website. I have been using this preliminary system to begin retraining horses having problems for about 15 years now and find it extremely reliable. We do have to learn to think in a more structured and specific way to apply ES, but the improvements it offers in the way of equine welfare, well-being and human safety are immense.

IN A NUTSHELL

ES equine learning theory helps us to train, ride and handle horses in ways they are mentally able to understand. Some of its principles are established classical principles as well. We have discussed the concepts of dominance and exploitation in equestrianism and concluded that dominance is inappropriate but that exploitation is unavoidable if we are going to use horses for our own purposes. Therefore, ethically, we must do so in as humane a way as possible and be ready to abandon principles and techniques which do not promote or enable the horse's welfare.

Negative and positive reinforcement are explained here, and we have discussed the training scales of the FEI and that devised by ISES, which is strongly recommended because of its logical structure. We also talk about overshadowing—when one stimulus is stronger than a simultaneous one, it is said to 'overshadow' the latter, which can be a confusing process for the horse. Overshadowing is a double-edged sword because it can be used also for good purposes such as habituating a horse to, say, clippers, so overcoming his fear.

Early-days training of foals in leading and generally becoming confident and developing some independence is covered, as is backing using different approaches.

The good and not-so-good points of each method are considered. Bitting is also described, and the old practice of leaving a young horse bitted and strapped up and tethered in his stable is denigrated as being cruel and dangerous.

Backing and riding away in different methods are covered along with lungeing and long-reining, and ES foundation training is described in some detail. Training aids receive due attention and in general are neither favoured nor felt necessary by the author.

8 On the Bit—The Key to the Door

We hear a lot about 'on the bit' in riding but it seems that quite a lot of people do not really understand what it means and, when they do, how to achieve it and why they need to. It represents a milestone reached in the training of a compliant, balanced riding horse with what are often called 'light sides' and a 'soft mouth'. These latter phrases just mean that the horse responds quickly to light aids of legs/spurs, whip and bit. They were prized qualities until a few decades ago and still are in authentic classical riding and Equitation Science because they indicate a horse who has adjusted his balance naturally and independently to cope with his (judicious) work and so developed his posture because of it. He is ridden humanely and effectively with due regard to the kind of animal he is.

The early work discussed in Chapter 7 will have brought the horse to the point at which he is readily responding to our aids under saddle to go, slow down and stop, and turn the forehand and hindquarters both ways. He will step back and forward, and turn, lightly in hand and remain calm. He should now be ready to progress to actually going in horizontal balance (*see* next) on the bit.

WHAT ACTUALLY DOES 'ON THE BIT' MEAN?

What it does *not* mean is a horse who is leaning on or pushing into his bit, which could be the impression it gives. 'On the bit' describes a horse who is going in what I call 'self-balance' rather than, at this stage, the more advanced classical self-carriage. He is able to respond to light aids and maintain his own balance under a good, balanced rider, also self-maintaining his gait, speed/tempo, rhythm and resultant posture (about which more shortly). In ES terms as well, he is capable of all these qualities.

The 'rounded' or flexed position of on the bit, which is believed to be needed in training but is actually an incidental by-product *of* training in classical riding and any biomechanically correct training, describes the horse as being flexed at the poll where the back of his skull meets the first vertebra of his neck (the atlanto-occipital joint) so that the front of his face (the nasal planum) is about 12 degrees *in front* of the vertical at walk and six degrees in faster gaits. The head and neck, though, are not all there is to it. Like full self-carriage, on the bit is a state of being that involves the whole body.

BITS, NOSEBANDS AND CONTACT

If we're talking about on the bit, it would be good to know what level of contact constitutes that, neither too little nor, as is the most common today, too much.

DOI: 10.1201/9781003121190-8

BITS

Bits were obviously invented thousands of years ago to help riders, often stirrup-less and saddle-less, to control their horses, and there are some cringingly awful examples out there to give us and our horses nightmares. The best thing I can say about any bit is that it should be comfortable for the horse. It seems obvious, then, that a bit a dismounted horse mouths gently, or not at all if he's standing still, but does not champ, try to get his tongue over or try to get rid of, is the best choice for him or her. If there are then any adverse reactions to it when the horse is ridden, it is obvious then that they are the fault of the rider and/or his or her trainer, or perhaps whoever tacked up the horse.

If I tried a bit on my old Thoroughbred, Sarah, that she did not like, she would stand with her head low and vertical to the ground, her mouth wide open, and shake it till her humble servant removed it.

My trainer mentioned earlier, Dési, would not have a double bridle on his place— 'Too much of a mouthful for any poor horse', he explained. All his horses went beautifully in half-moon pelhams and the younger ones, or those in for rehabilitation or schooling, initially wore shaped, double-jointed snaffles with a central plate angled to lie flat on the tongue (not Dr Bristols) as being the closest one could get, back in the 80s, to a bit that moulded to the horse's lower jaw and tongue. Today my favourite snaffle is an eggbutt in a lozenge design with shaped cannons.

There is something about a simple, half-moon (not arched) pelham bit. I have found *many* horses go really kindly in them and seem more comfortable than in a snaffle. It seems to me that something about the way they hang and their weight distribution and feel in the mouth causes the horse to naturally slightly flex at the poll and 'give' comfortably, even when dismounted. A half-moon snaffle doesn't have the same effect.

I prefer close-linked, metal curb chains to the coarser-linked ones as giving a smoother feel and movement in the curb/chin groove. I also do not like leather or elastic curb straps as, in my experience, they just will not lie down comfortably in the curb/chin groove, even with a lip-strap, when the curb rein is used, but slide up the lower jaw bones where the skin is very thin, causing discomfort.

NOSEBANDS

Nosebands can greatly interfere with the use and effect of the bit. While doing some research for a *Tracking-up* magazine article about them, I came across a paper by an American researcher who wrote: 'If a horse cannot yawn with his noseband fastened, it is too tight'. Point taken indeed.

The International Society for Equitation Science (ISES) has available a wedge-shaped taper gauge which is slid between the noseband and the nasal planum and which is marked to show the humane level of tightness. Finger width has long been used to check nosebands for appropriate, humane tightness and the recommendation is that we should be able to fit the sideways width of two fingers between the noseband and the nasal planum, some say three fingers. For your interest, it seems from

Figure 8.1 This illustration is of the late Miss Sylvia Stanier, LVO, riding Mary Chipperfield's Andalusian stallion, Pedro, in a High School display at Dublin Indoor International Show, 1982. Note the snaffle bridle.

some studies that, although we know that tight nosebands can cause mouth injuries, from bruising to blood, working a horse *without* a noseband can also cause mouth injuries, so perhaps the two or three fingers check, or a proper taper gauge, are our, or rather our horses', best options. If only this could be made an international rule— *and implemented.*

LEVELS OF BIT CONTACT

The ES contact scale I use runs from 0, meaning a loose rein with no direct bit contact at all, to 10, meaning as hard as you could possibly pull in an emergency. A light contact is considered to be from 3 to just over 4, although I have known plenty of horses who were uncomfortable with that and have been fine, responsive and on the bit on less than 3. I have seen a scale running from 0 to 5, but I think the 0–10 scale gives more scope if you want to be precise.

An ES 'neutral' contact, which I call an 'in-touch' contact, is just enough to keep your reins straight from bit to hand without pulling back the corners of the horse's mouth. (Of course, a good snaffle-bit height is for the bit to create no more than one wrinkle at the corners of the horse's mouth; pelham bits should just touch them without creating any wrinkles.) The contact scales previously mentioned relate to snaffles and work well for pelhams via the bridoon rein, or double bridle bridoons.

Of course, curb bits, with their potential to greatly intensify pressure, would not, I hope, be used according to the scale except perhaps on a level of one—a suggestion to the horse to relax or 'give' his lower jaw and flex to the bit.

I read a study some years ago which found that if the cheeks of a pelham or double bridle curb are pulled back to the accepted 45° angle with the line of the lips, the amount of pressure that would create in the mouth is enough to cause considerable pain. (Remember that the tongue and lower jaw are held in a threatening vice between the bit mouthpiece and the curb chain.) This would be with the curb chain fitted as correct, so that it fits down into the curb/chin groove and stays there in use, not above it and not down on the chin, and one finger can be slid easily along the groove under the chain.

Any bit involving a curb action also needs to be treated with great caution. In fact, the need to 'bit up' a horse in a strong bit shows that his foundation responses are not in place; in other words, his schooling/training or riding leave a lot to be desired.

The argument in favour of using double bridles is that they permit a high level of finesse and demonstrate the skill of the rider in using them and that the horse is schooled to a level at which he responds 'obediently' to them and understands their action. In fact, a good rider can certainly get as good a result from a pelham as from a double bridle and, indeed, from a simple snaffle. The sign of a good horseman or woman is that he or she uses the simplest equipment and the most comfortable for the horse, plus correct, ethical training and riding, to get their results.

Dr Sharon Cregier, who kindly wrote the Foreword to this book, has said: 'My main criteria of good horsemanship are, the less equipment on the horse, the better the rider, and, does the horse look forward to his rider's visits?' A good example of this is our photograph of Sylvia Stanier and Pedro giving a High School display on page 123.

Two traditional descriptions of a good contact are that you should hold your reins as though you are holding a small bird in your hand, securely enough to keep it there but not to hurt it, and also to imagine the 'contact' you would need to hold the hand of an obedient, small child when crossing the road, enough to keep him or her secure without hurting their hand.

The quality of your contact also makes a big difference to your horse. Many people think that riding mainly on a loose rein, out of touch, is kind, but there are two main reasons why it isn't unless your horse is in true self-carriage on the weight of the rein:

Figure 8.2 Progressing on the lunge. This rider has a good seat and posture, his legs are well dropped and he has a sensitive hold of the reins, having progressed from the initial stages of lunge work. The horse is Anne Wilson's mare, Secret.

1 Riding mainly on a *loose* rein, apart from giving your horse a break, does not encourage the horse to use his body appropriately to strengthen it for weight-carrying. His back will forever sag, and he will never come on to the bit or strengthen his 'driving force'—his hindquarters. This means he is going in a way, carrying weight extra to his own, that is stressful to his body and can cause physical injuries, particularly to his back and possibly his forelegs due to two-thirds of that imposed weight being on his forehand.
2 A loose rein does not allow for a gentle but meaningful contact. If the rein is only *just* loose, what I call 'flapping', the horse, particularly in trot, will be feeling a jab in his mouth twice in every stride which must be really unpleasant—imagine it. It is a sure-fire way of encouraging him to go defensively—like a banana with his head up and nose poked out, his back down and vulnerable to the rider's weight, and his hind legs strung out behind him, doing no useful work apart from stopping himself falling over.

A light but clear and definite contact is what's needed. Trained horses will treat this as a useful parameter. Try your horse on a 3 contact (on the 0–10 scale) which is plenty for general hacking and riding around, and see how he seems to like it; then vary it until he seems comfortable, confident and trusting his bit, going up to it gladly but not leaning on or avoiding it. And make sure your reins are not flapping or that you are not pulling. If your horse starts to feel heavy, try vibrating the aid quickly and

Figure 8.3 Many problems in riding occur simply because the horse's tack is uncomfortable. This bridle leaves plenty of room around the ears, the throatlatch is fitted correctly loosely, the noseband is halfway between the corners of the horse's lips and the sharp facial bone so it will not rub there, and the bit is *not* pulling up the corners of the horse's mouth.

gently (see next paragraph), which often works well. It is kinder to give your horse a contact he can trust than to ride without one (if he's not ready for that) or, worse, a halfway one.

Vibrating your aid, quickly, gently but definitely, whether leading in a headcollar or riding in a bridle, is very attention-grabbing to the horse and much better than increasing the pressure too much, but there is a way to do it. If your horse does not respond to a light aid, increase the pressure *slightly*; then, if it's still not working, rather than increasing it more, quickly vibrate your contact but with an on-ON-on-ON feel rather than an on-off-on-off, jabbing feel which is most unpleasant and won't put you in your horse's good books. So, you keep in touch with your contact, but quickly and rhythmically vary the pressure, as described and if needed.

At the opposite end of the scale, it is also very easy for conventional riders to fall for the modern theory that your horse needs to be held in firmly all the time, to develop his physique and enable him to bring his back end under, or to help him balance. This is not how equine biomechanics work, as already described. Conversely, a general, in-touch contact gives your horse confidence, develops his balance under weight and does encourage him to carry himself correctly. It also gives you immediate control if you need it. With a heavy, sustained contact, a rider could actually be abusing his or her horse.

There is also the viewpoint that you should give (lengthen) the rein if your horse applies pressure to it. Again, this may not be helpful. It's tempting to let him take your hands forward, to straighten out your arms (dealt with later) and even to lean forward, losing your position. *Keep your position,* and if he really seems to need more rein, open your lower three fingers and, if necessary, lengthen the reins a little by letting them slide just a little through your thumbs and index fingers. It could be that his muscles are ready for a rest, in which case stand on a *free* rein, holding the reins by the buckle end and keeping your hands *still*. It is surprising how many riders don't seem able to do this but, if this is the case, it really must be practised because it is a valuable way of resting a horse physically and mentally and increasing his confidence.

Figure 8.4 The modern very firm and unrelenting bit contact does *not* help the horse balance or give him guidance, as is popularly believed. It results often in an unhappy horse, clearly shown here, with a shortened neck, a sagging back, an overbent posture and a too-low poll. It can cause considerable pain in the mouth and bit injuries.

RIDING FROM FRONT TO BACK AND VICE VERSA

RIDING FROM FRONT TO BACK

There is a very common misconception, held at, apparently, some of the 'highest' levels in equitation, that positioning the horse's body and particularly the neck and head in a 'rounded' posture by means of bit use or training aids will cause the horse to adjust his balance and weight-carriage rearward and will, how I do not understand, result in the horse being more responsive to his rider's various aids.

Anyone can pull in a horse's head so that the nasal planum is on, behind or just in front of the famous vertical line dropped from his forehead to the ground. 'Placing' the head significantly like this does not beneficially enhance a horse's way of going because he is working under coercion. It does not imply that the horse is on the bit or 'on the aids' (responding quickly and lightly), and it does not position the rest of his body in the rounded or bowed posture that facilitates weight-carrying, a raised forehand, lightness and agility.

If the rider pulls hard enough and sustains an unremitting, heavy bit contact, two sorts of faulty posture often result:

- The poll too low (not the highest point of the outline) and a behind-the-bit head carriage, which is the most common; or
- A high-headed carriage flexed at the poll, the nasal planum vertical but the neck kinked down at its base instead of being lifted.

Both can result in a dropped back, semi-functioning hindlegs and a shortened, cramped neck and throat. The horse may well transfer some weight to his hindquarters mainly to try to avoid his discomfort while trying to work in a contrived, *incorrect* posture which can only distress him, confuse him and cause him considerable discomfort.

To work in an overall 'rounded' or bowed posture from tail to poll, a horse needs to be strengthened and given time to adjust his own balance and carriage. In fact, it is the preparatory work and its correct, logical progression ('shaping' in ES parlance) that will produce naturally the sought-after posture or outline. Forcing it can result in these significant disadvantages:

- A contorted, back-down way of going.
- An exaggerated vertical or behind-the-vertical flexion at the poll, making it difficult for him to see where he is going unless the horse 'looks up' all the time, which is very uncomfortable.
- A pulled-in, shortened neck and cramped throat, creating difficulties in swallowing his own saliva and in breathing.
- Difficulty also in bringing the hindlegs forward under the torso to carry weight and propel the horse forward.

Such a horse cannot respond to aids either at all or, at least, in lightness, and must be very uncomfortable in his work. It also seems reasonable to suppose that riding from front to back using 'time-saving', forceful work can be responsible for much hindleg lameness in horses and general soft-tissue damage, including muscle damage in the forehand and back, plus mental health issues.

RIDING FROM BACK TO FRONT

Achieving strength, balance, posture and quick, light responses depends on riding from back to front. The work (mainly correct transitions and bending exercises) will

- First, strengthen the horse and familiarise him ('classically condition' him) to going in a certain way which, although he is carrying the weight of a rider and saddle, will not be onerous to him if it is done correctly and gradually increased in time and difficulty.
- Second, it will naturally encourage him to adjust his balance slightly to the rear, as mentioned. In brief:

1 He will slightly 'tuck his bottom under' or tilt the rear part of his pelvis down and forward, enabling his hindlegs (which are attached to the pelvis, of course, via the hip joints) to come a little more forward under his torso.
2 His spine will generally be slightly raised, increasing its natural, slight, upward bow or arch, which strengthens his back and increases its weight-carrying ability, an arch being a much stronger physical structure than a downward dip—think of bridges and buildings.
3 This enables him, in time, to bear the weight involved more easily.
4 His forehand will lift slightly and 'lighten', increasing his agility.
5 His neck initially will be *stretched forward and down* with a slight arch, his head coming into a position closer to the vertical and his poll about level with his withers, in a natural action to counterbalance the rearward shift of weight to his hindquarters. (You can try this yourself even though we are vertical and our horses horizontal. Stand up straight with your feet pointing forward and directly beneath your hip joints. Keep your arms by your sides and maintain controlled relaxation, to which you are, by now, no stranger. Now slowly and carefully lean backwards. You will almost certainly find that your head tilts forwards to counteract your change of balance.)
6 Finally, your horse will, if the work continues, develop the *upward, forward and rounded stretch* and carriage of an advanced horse, with his nasal planum on or preferably just in front of the vertical.

This posture is maintained without stiffness on the part of the horse. Like a good classical rider, he works in a state of controlled relaxation, neither stiff nor floppy. The back and torso undulate slightly, vertically and laterally, the ribcage and abdomen swing from side to side a little, according to the gait and the *dock*, not just the tail hair below it, swings gently from side to side indicating relaxation but, overall, the posture and action previously described are voluntarily and naturally achieved and maintained by the horse throughout his work—not because the rider is 'holding him together'.

When the horse is on the bit, his contact will probably be somewhere around point 3 on the 0–10 scale. He will still be mainly in horizontal balance, under 'stimulus control' of his rider (that is, he will obey aids/stimuli from the rider reliably, quickly and lightly) and the feeling the rider starts to get will leave him or her in no doubt about the correctness and value of being patient and applying correct training to his or her equine partner. It is rather like driving a powerful speedboat or sports car. You feel as though you are being pushed forward and up from behind (you are, by the hindquarters and legs), and you will never again look for an excuse to not ride! That particular horse, anyway.

Part of a good physical development and care programme for a working horse involves appropriate stretches. Everyone knows how to do a 'carrot stretch' (which can be overdone) but few know how correctly to stretch the neck, legs and back. Stretching helps to complement the contracting action of the muscles in work and promotes good muscle health. It is well worth booking an appointment with a physiotherapist or other qualified, registered bodywork practitioner, to learn the right way to carry out the most appropriate stretches for your horse and how to give him a good, basic massage every week. Regular physical check-ups from such a therapist are recommended, as well as your normal veterinary health checks.

THE BALANCE OF HORSE AND RIDER

To understand 'on the bit', it is necessary to appreciate the different main states of balance of a horse and his rider, and the importance of the correct fit of saddle and girth.

Naturally, horses carry a bit less than two-thirds of their weight on their forehands, the forehand being mainly for weight-bearing and the hindquarters mainly for propulsion. In a horse being strengthened to carry the weight of a rider, we seek to get him ultimately to transfer some of his weight (including that of the rider and saddle) back a little towards his hindquarters, so that his centre of mass or gravity is as far as possible directly under that of the rider. As already mentioned, the horse's centre of mass is about a hand's breadth behind the withers, two-thirds of the way down inside his thorax, directly under his spine. Taking his weight back a little not only takes weight off the horse's forelegs but also lightens and actually lifts the forehand slightly with work, helping to make the horse 'light in hand' to bit aids/cues and increasing his overall agility.

For practical purposes, we can take the rider's centre of mass/gravity to be the seat-bones, although it is actually slightly above them. In a modern flatwork saddle, often a dressage saddle, one that is balanced and fitted by a professional, qualified saddle fitter to the horse's conformation and action, and also to the rider's make and shape, the seat-bones will be central in the deepest part of the saddle so will probably be positioned slightly back from the horse's centre of mass. As the horse will be taking his centre of mass back slightly, ideally ultimately to be as closely as

Figure 8.5 Riding constantly on a too-loose rein most of the time might seem kind
but it is not. This rider's straight arms indicate that perhaps she feels she
is being considerate to her horse. In classical riding and ES practice, too,
it is advisable to give the horse a gentle, in-touch parameter with the bit,
so that it is used mutually as a means of communication. A loose rein,
apart from a rest during work in halt or walk, does not help the horse
to come into a posture that is going to help him strengthen and round
up over time, and get his weight back with correct work. In fact, it puts
considerable strain on the back, which, under the weight of a rider, will
naturally dip and become painful and possibly injured. Also, riding on
a rein that is *just* too loose results in the contact being erratic so that the
horse gets a little jab in the mouth at every step or stride, resulting in a
poked nose and, again, a dipped back and disengaged hind legs.

possible directly under that of the rider's seat-bones, both centres will be then largely
harmonised on the same vertical plane. This makes carrying the rider as effortless
as possible.

THE SADDLE AND GIRTH

Both our and our horse's centres of mass shift slightly in movement—the horse's
mainly forward and back and ours mainly up and down. The faster a horse goes the
more forward is his centre of mass. In fast cantering, galloping and jumping, the rider
will adopt a forward, light seat with his or her seat just brushing or out of the saddle.
This means most weight will be taken on the stirrups. Contrary to popular belief, this
does not free the back from the stress of moving weight because the rider's weight trav-
els up the stirrup leathers to the stirrup bars (which are positioned somewhat forward of
the deepest part of the saddle seat) and from there down on to the horse's back.

So, in a normally well-fitting saddle, the rider's weight in a forward or light seat is concentrated on the horse's back in the areas under the stirrup bars. Endurance riders who think riding standing in the stirrups relieves the horse's back might bear this in mind. It is important, therefore, for all riders to learn to take some weight on the inner thighs and knees to help to spread the load.

Putting a saddle on too far forward, as is common, does a horse no favours; quite the opposite, in fact. Often done with the best intentions, to position the rider's weight where it is more easily carried, it actually makes matters worse. Here's why:

- Every time each foreleg moves forward in the air in what is termed its 'swing phase' or 'in protraction', the top of the shoulder blade moves back and if the saddle is too far forward the two impact on each other at every stride, so the horse's shoulders can become bruised and sore.
- When each foreleg moves backward in its 'stance phase' (the hoof being on the ground and the foreleg moving backward over it 'in retraction'), the girth, necessarily being too far forward because of the saddle placing, can, and usually does, dig into the space between the elbow and the ribcage, causing pain and bruising in that area, too.
- Being too far forward can unbalance the saddle, tilting it up in front so the rider's weight drops back and is concentrated too near the cantle, not far from the most sensitive part of the back—the loins or lumbar area.
- If placing the saddle correctly means the rear part of it contacts the loin area, the saddle is too long.
- You should always be able to fit the width of the side of your hand between the front of the saddle and the top edge of the shoulder blade, by the withers, when the horse is standing still, and the width of your flat hand between the point of the elbow and the girth when the saddle is girthed up.

Finding a saddle that fits the horse underneath and the rider on top can be very tricky, which is why the services of a professional, qualified saddle *fitter* are recommended.

Girths with elastic inserts at both ends are recommended, to allow for some 'give' when the ribcage expands with each breath: if only one end is elasticated, the saddle is pulled over to the opposite side as the horse breathes, causing friction from saddle and girth and uneven weight distribution. If no elastic inserts are used, the girth can exert considerable pressure to the point of restricting breathing (because many people girth up too tightly these days) and causing great discomfort.

STATES OF EQUINE BALANCE

For the sake of simplicity, we'll consider the three main states of a horse's balance and how they affect him and his rider. (*See* diagram on p. 156).

On the Forehand

Most riders will be familiar with this natural state of the horse. We have already said that horses carry roughly two-thirds of their weight on their forehand. If a horse has the conformation of the highest points of the withers and the croup being the same

Figure 8.6 This rider is assisting her very novice horse with her exemplary position and in-touch contact. The horse is walking along carefully, balancing himself and his rider, and is calm and interested.

height, this is fine and is the usual condition of most horses. Carrying a rider, for a novice horse, one out of condition or one having been poorly trained, can make his forehand that bit more 'loaded' and decrease his agility, but we allow for this and athletic training will gradually improve it.

If a horse has croup-high conformation, which often seems to accompany a rather poor front but not always, he will give the rider a definite 'downhill' feel which can be particularly disconcerting when actually riding downhill. Such horses often have a naturally low head carriage which adds to the feeling of insecurity. The good news is that these horses' way of going can be greatly improved as their muscles develop, they accept a reasonable bit contact, learn their responses and go habitually (usually their choice) with their heads a little higher. The tendency to go with too low a head carriage fades as the topline muscles become stronger with correct work, but such a horse may always need to rest his neck muscles more than other horses.

Horizontal Balance

This develops when a horse has been in training for, variably, some months or a year or so, and is the type of balance in which going 'on the bit' can be achieved. The horse feels 'even' to ride and will have been partially strengthened with mainly bending and transition work to enable him *naturally* to take a little weight back towards his hindquarters. As always, the aim in classical and ES riding for light responses to go, stop/slow down and turn will have been achieved and become habitual, rider permitting.

Advanced Balance

That description is for want of a standard expression. It means that a horse is readily taking his balance back, consequently lightening his forehand, and will be on the bit and ready to progress (*see* Chapter 9) to more advanced work such as collection followed by extension, and going in the ultimate self-carriage on the weight of the rein. Going on the bit will be habitual to him by now provided he is ridden well.

PREPARATION FOR 'ON THE BIT'

Because being on the bit is a whole-body condition and not merely a case of the neck being 'rounded' and the head carried as previously described, we need to know what to establish in the horse before it is possible. The three ES requirements or precursors for on the bit are:

1 Longitudinal flexion.
2 Lateral flexion.
3 Vertical flexion.

1 Longitudinal Flexion

In longitudinal flexion, the horse will bring into effect his 'vertebral bow' in that he will go (as previously described) from his hindquarters which are slightly tilted or 'tucked' under, enabling his hindlegs to work a little more forward under his torso. This causes his spine to bow upwards a little, increasing a little its natural, slight upward arch, particularly in the suspension phases of trot and canter and, importantly, his neck is stretched forward and lowered (again as previously described) so that he carries his head about level with his withers, and his nasal planum is carried closer to the vertical depending on his gait.

2 Lateral Flexion

Here, the horse turns his head at the poll/atlanto-occipital joint and, strictly, the top three neck or cervical vertebrae in the direction of any turn the rider requests. Green and free horses do not do this: it occurs with training very gradually and is a requirement, particularly in dressage, of the trained horse, requested by the rider in turns, corners, circles, loops and serpentines, and in lateral exercises requiring bend. It is usual, on curved lines, to aim for the degree of lateral flexion to match the arc of the line. In regard to recent research findings, it is recommended now that horses are not asked to perform circles or curves of less than 6m in diameter.

3 Vertical Flexion

This is the flexion at the poll joint of the head in and down while maintaining the poll as the highest point of the horse's outline and the nasal planum *not* behind the vertical (degrees of vertical flexion were given earlier). Many people overdo vertical flexion, yet it is a dressage rule that the poll must be the highest point and the horse not BTV—behind the vertical. It is sad that overbent horses often get high dressage marks while their riders and the judges seem to disregard these requirements. I understand that, not surprisingly, this posture can adversely biomechanically affect the horse's body, so judging criteria must surely be reconfirmed and adhered to.

Bend

This is another feature of a trained horse which is involved very much with flexion. 'Bend' is a lateral curving of the vertebral column or spine in accordance with following any curved line or all lateral movements except leg-yield and is a feature of a well-schooled horse. Because the horse's spine cannot curve uniformly, as explained earlier, we need to get rid of our traditional idea of the spine following evenly the line of a circle. We can create the illusion with moderate lateral flexion.

Bend is considered desirable because, from our viewpoint, it looks more beautiful and it enhances a horse's suppleness. It is said that bend is what enables a horse to carry his hind feet exactly on the tracks of his forefeet on curved lines. However, if you watch any non-bent horse, either ridden, in hand or free, going on any curved line but a very tight one, you will see that he does this anyway.

I learnt to ride partly on a beach, and we all had great fun competing with each other at recognising the tracks horses made in their various gaits, including horses who had gone before us that day as well as our 'own' ponies. As our expertise grew with us, we started thinking about flexion and bend, but found that, although it seemed to make our ponies easier to handle and better balanced on curves, it made not a jot of difference to the positions of their hoof prints in the sand, the best time to see this being about an hour and longer after the tide had gone out.

Now, sensibly, circles of less than 6m in diameter are advised against for horses due to the now proven lack of *overall* lateral flexibility of the horse's spine, although I suppose they would still be all right for ponies.

THE RIDER—SEAT AND TECHNIQUE

HAPPY MEMORIES AND LEARNING THE HARD WAY

Our riding master at the small riding school I attended in my childhood and teens was always saying 'Sit up, sit down and sit still' which we knew meant precisely—(a) sit up with control but not stiffness, (b) relax your seat and legs so you could drop down in the saddle from below the waist and stay there, and (c) sit still apart from moving with your pony and to give aids because every movement we made in the saddle could be interpreted by our ponies as a seat aid. If we slopped from side to side, our pony would move from side to side with us under our weight to help him stay balanced (illustrated by our riding master on the sand, and copied by us in case we didn't believe him!), so it was our aim not to do that but stay central, unless we wanted to turn, By the time we progressed to larger ponies and horses, this had all become second nature. The poem we all had to learn and recite regularly was:

> *Your head and your heart, keep up.*
> *Your hands and your heels, keep down.*
> *Your knees keep close to your horse's sides*
> *And your elbows close to your own.*

Excellent advice for any rider, especially in the days when we weren't allowed to 'dismount without permission' (fall off, which attracted a forfeit), and we used to ride, race on the beach and jump, including bareback, in hats made of felt and 'secured' by a piece of elastic under our chins. Our riding master rode in a trilby and even those regular visitors (with their own posh horses and ponies) to our small yard of former railway-horse stables, who came for his excellent instruction (he being an old-school, classically trained, former cavalry instructor), had 'hard' hats made of cardboard covered with velvet and a silky, lightly-padded lining, but still with the elastic band.

Usually, we had to change horses and ponies at half-time, after giving our mounts a ten-minute break with loosened girths and run-up stirrups, during which we had a stable management lesson. The visitors were glad to ride the school horses (who were very much better behaved than most of theirs) and we got to ride theirs for a change, my favourite being a gorgeous, springy Arab mare called Tanner (meaning Sixpence—2½p today).

After the retirement of our riding master, I could not find a riding school to compare with his but kept his principles close to my heart and got to ride many lovely horses and ponies who responded really well, not to me so much as the way I rode. Then in the 80s, having recently lost my own beloved first horse, I received a phone call from a Captain Dési Lorent, who had seen an advertisement for the magazine I was self-publishing then—*EQUI*. He asked me, in his strong, French Belgian accent, if I would be interested in having a long weekend's free tuition at his yard in Devon, where he taught Nuno Oliveira's way of riding and schooling, being a long-term friend and student of his—20 years long, in fact. There was method in his madness, because he wanted me to write an article about his classical training yard in *EQUI*. Free publicity for him, free jaunt for me. Would I just! I ended up going back to my childhood and then some, and studied with him, very part-time, for two years, after which he became disenchanted with Britain and left.

Studying with Dési was like coming home, with bells on and flags out. I had been to various different conventional riding schools and centres in the intervening years and constantly been told that I was old-fashioned and 'we don't do that now'. The trouble was, I didn't like what they did do, and their horses didn't seem to like it, either. I jumped at going to Devon whenever I could (no longer free of charge, of course!). The way Dési taught the classical seat is what I shall describe here, and our voluntary homework was all of Oliveira's books, plus others'.

THE 'CLASSICAL SEAT'

There is a lot of misunderstanding of the classical seat (*see* p.109) from those who ride conventionally and/or have never had real classical tuition. Good classical riders look still, immovable and, to the uninitiated, can seem stiff, rigid and perhaps a little bit snooty! However, many onlookers comment on the fact that they can't be seen to be doing anything yet the horses perform like a dream. In practice, the classical seat is all about good, healthy and balanced posture, controlled relaxation, common horse sense and putting the horse first—this latter point being why classical riders sit as they do, so that they can really *partner* their horses, disturbing their balance as little as possible while going with their movements and absorbing them, delivering their aids/cues/signals, and having due regard for their horses' mental functioning and capacities.

I once saw a clip on YouTube of Masai warriors wandering their ancestral lands in Africa, and their posture was stunning—upright, proud, relaxed, natural and so impressive. 'That's the ticket', I thought. As a teacher, those I have found easiest to teach have been dancers, especially classical ballet dancers, skiers, skaters and gymnasts, who all understand the essentials of posture.

CHEZ CAPITAINE LORENT

BACK TO BASICS

So, at Dési's, newcomers, whoever they were almost without exception, were put on the lunge with no reins or stirrups but a neckstrap for a comfort blanket. The very first thing to be taught was to sit stretched up from the waist, shoulders rolled up, back and pushed down, and gently but definitely kept there. Your neck you pushed back into your collar (real or imaginary) and you were often asked to lower your chin. Dési placed great emphasis on keeping your elbows '*IN PLAZE*', which meant your upper arms *vertical* most of the time and your elbows, therefore, in place *at the sides of your hips*. This, together with your stretched-up torso, stabilises your hands without imprisoning them.

Your body below the waist was treated very differently. You had to aim to sit on your seat-bones as default, keeping one on each side of your horse's spine: obviously, this involved imagination but makes sense in practice because it immediately improves your position and your balance. Those who were a little hollow backed, like me, were told to *slightly* tuck our bottoms under, just a fraction, which surprisingly made our seat-bones much more obvious to us.

THE SECRET OF IT ALL

The next part is the Big Secret of the school of classical riding Dési taught, learnt very largely from Nuno Oliveira, and which I have never heard or experienced anyone else teaching in quite this way—and it's not easy at first. Remember, we had no stirrups or reins on the lunge. He used to exhort us to 'make your seat and your legs like your great-great-grandfather's—dead'. This meant, in no uncertain terms, *TOTAL* relaxation of seat and leg muscles so that you were, in effect, balancing on

Figure 8.7 A novice horse walking well, in front of the vertical with a natural length of neck due to not being held in by his rider.

your seat-bones with your legs dangling lifelessly down your horse's sides, hands on your thighs, toes hanging down.

In conventional lungeing, riders are normally told to raise their toes and drop their weight through low heels, but actively raising the toes involves contracting the muscles on each side of your shins, which stiffens your legs—exactly what we don't want. The lunge was the ideal place for practising this relaxation because we had no need to use our legs to aid the horse.

It was amazing how difficult it was to really, fully and completely relax our seat and leg muscles. Just when you thought you'd got it, still on the lunge without stirrups, Dési would put a hand between your lower leg and your horse, pull it away a few inches sideways and let it go: if it flopped lifelessly back you had, indeed, got it, but if it stayed out or slowly returned to the horse's side it was because your leg muscles were in tension, controlling the leg, and you *hadn't* got it.

We mastered it eventually, not least to escape a telling-off (to use a polite word) from Dési, whose bellowing (in French if he was really angry) could be heard in the village down the hill. (We took it in turns to go down for the shopping every day, and the owner of the mini-mart used to ask: 'Who was it getting it this morning?') You either collapsed in tears (including some men) or gritted your teeth and determined not to give him anything to roast you about. Fortunately, with my background of military-style classical instruction as a child, I was able to do that. Having had ballet training up to my teens also meant that he couldn't complain about my posture (apart from my having consciously to turn my feet in)—but be assured, ballet training is not an essential precursor to learning to ride in the classical seat.

Once we were able to walk, trot and canter with our legs dangling like rubber and our seats perfectly balanced in the saddle, we were allowed stirrups (but still no reins because of the sacredness of horses' mouths), which felt really weird by then. We all realised that our stirrups had to be longer than previously. The lower-body relaxation and seat-bone positioning were aimed at getting us to drop and mould around our saddles which gave us a terrific sense of security. The elbows-in-plaze was easy enough for me but a lot of students taught the conventional way had trouble breaking the habit of dropping forward with rounded shoulders, hollowed chest and straightened elbows, and giving too much with the arms, occasioning Dési's ire (*'you abandon my 'orse!'*) with humiliating regularity. We had to sit up, shoulders firmly but gently held back and down in controlled relaxation, and mostly just opening our fingers to lengthen the reins when needed. No wonder he got through so many Havanas and so much Dimple.

On Trial

Once he thought you were fit to let loose again, it was still in the indoor school (converted from a barn) in case you upset one of his horses. It was the old classical size of 30m by 15m (there was also an outdoor manège of 20m by 40m) and had lovely diagrams and pictures on the walls of classical positioning and aids, which I believe were done by one of his students, and which he used for instilling everything into our overloaded brains. We knew we were on trial, but by then we had realised how much better it was riding this way (even better than my childhood riding school which had, though, instilled the basics and more into me) and were besotted with it all, loving every minute of it—almost. Several conventionally trained instructors visited and said they could not go back to their old ways. Some others left, totally fazed, after a day or two.

Our first test was to walk our horse round the school on the outside track. Simple enough, one would think. Dési was always reminding us that 'where you look, and where you put your weight, your horse will go', which I had learnt as a child, so we looked ahead where we were going and, to turn, moved our inside seat-bone forward and weighted it a little by stretching down our inside leg, dropping the weight through our heel (without letting our lower leg move forward) but never leaning our upper body over. (Basically, our torso had/has to be kept upright to help the horse balance us and stay upright under our top-heavy weight on his back.) That was all at first. The horses all knew this and if we didn't do it, thinking they would just carry on round a corner, they stopped, head to the wall, rather than turning unbidden like 'ordinary' horses. Dési was waiting for this and would call out: 'Why do you stop? *Marchez!* I thought you said you could ride'. So encouraging!

To walk our horse forward, we learnt to give an *inward* squeeze with our two legs immediately behind the girth for 'go'. To slow down or stop, we had to sit up and lighten our seat and *then* feel the bit, both of these moves without legs because a squeeze from the legs meant/means 'go', the opposite to what we wanted. If we wanted rein-back, from halt of course, we would again lighten our seat and give a feel on the bit—no leg aid as such, just lower legs slightly back in a very light 'position aid' to help keep the quarters straight. If you are sitting centrally, anyway, even this should not be necessary.

Figure 8.8 A schooled horse trotting well, on the bit with natural posture of his head and neck due to no restrictions from tack or rider, and having been developed with correct strengthening and lightening work. The rider is in a soft sitting trot.

It is, as you'll know by now, having read this far, a basic principle of ES not to give opposing aids at the same moment, as is usually done in conventional riding under the guise of, for instance, the confusing 'driving your horse up to the bit' and 'riding forward into halt' techniques. We were taught to gently keep asking till the horse complied, then to stop the aid immediately—classic ES teaching. 'Don't keep asking for the salt once you've got it', explained Dési.

The importance of moving our seat, or rather allowing our seat to move, with the horse's movement in each gait was drummed into us continually. To do this, you needed to keep thinking of your body in two parts, that above the waist and that below it, as we had learnt on the lunge. Once you can do this, it's amazing how many 'difficult' horses relax under you and seem to wait happily for your aids, or signals, or cues, and how easily and readily they comply with your requests.

TABLE-TOP TACTICS—WALK

Think of your horse's back as a table with four legs. In walk, therefore, when the horse lifts, say, his left hind foot to swing it forward, the left side of your horse's back/the table loses its support and dips a little. With your beautifully loose but controlled seat and legs, you will feel this under your left seat-bone so, *keeping your body above the waist upright and still*, you can let your left seat-bone follow the dip down, then do the same on the right side. As the left hind meets and pushes against the ground again for its stance phase, your left seat-bone will be gently pushed up while the right one is dipping, only to be pushed up similarly as the right hind pushes you up and the left dips again.

So, your seat is alternately dipping left, right, left right, as your horse's back mirrors the swing and stance phases of his hind legs. This is much easier to learn without stirrups, but it's perfectly possible with stirrups once you get used to staying relaxed and just letting your stirrups take the weight of your legs. The important thing is to let your loose seat passively dip and rise with your horse's action, and not to actively do it yourself. This keeps you moulded around the saddle.

Our next test at Dési's, still in walk, was to be able to call out which hind foot was in the air, to check that our seat was suitably loose, and therefore sensitive enough to tell us this. So, we would be going along calling out 'left, right, left, right' in time with the dips, as the hind feet lifted in turn. This can be almost impossible for any rider not taught the importance of a relaxed, moulding seat and legs. You need a friend or clued-up teacher on the ground to tell you if you're getting it right or wrong—and, if wrong, it's because your seat isn't loose enough—even though you might think it is!

It might help you to know that the horse's ribcage and belly swing gently from side to side *away* from the dip in walk, so you could feel, say, the right side of your horse's belly slightly swinging right and pushing against your right leg as the left hind lifts, and vice versa. Do not let this movement cause you to swing your upper body from side to side! Remember, it is 'separate' from your lower body which is going with the movement, so stay stretched gently but definitely *up* and out of it. It is your seat that follows the dips, not your shoulders, which should remain level at all times. The stretching up from the waist and dropping down from the waist is the key to learning to think in this way. You are two halves joined loosely and flexibly by your waist.

Trot

Now—sitting and rising trot. In the *classical rising trot*, you are never as far from your horse's back as in a conventional rising trot and you can see from your horse's forelegs moving forward which diagonal you need to be on—normally sitting when you see the right foreleg swing forward if you are going to be on the right rein, and vice versa. It is the *sitting trot* in which you call out the dips and which can be tricky because you can't usually cheat by watching the forelegs from your upright, sitting trot position.

We learnt at Dési's (who I think would have been most interested in and accepting of ES) to feel the dips in sitting trot, too, which is easier because they are more marked, but quicker than in walk. It is also in sitting trot that you might start to tense up your seat and legs again, especially if you have no stirrups, although it is without stirrups that you can most easily feel the dips, so make an extra effort to keep your seat and legs loose. This is when you will really start to feel its benefits.

To start trying to call out the dips in sitting trot, start in nothing more than a jog or the ES 'little trot' (see Chapter 9) and see how you get on, then move up to an active trot just a little short of a working trot. The ES leg aids are really good because they are precise and clear, so move up to this trot with one inward squeeze of the upper inside of your calves, maintaining it till your horse trots, if necessary giving two whip-taps behind your inside leg, although I use an inward vibration with my lower legs to good effect. Sit as you did on the lunge and let your seat dip and tell you which hind foot is in the air and which on the ground. Keep your seat and legs loose or you won't succeed, and don't forget to keep your consistent, gentle contact, which is easy to forget when you are concentrating on your seat. Come down to walk with a single gentle but clear feel on the bit till your horse starts to walk, then release the aid, and call out the dips in walk, too. Then up to trot again on the other rein. You'll soon get the hang of it.

Bouncing up and down in sitting trot happens because the rider does not have properly relaxed seat and leg muscles, and you see this at all levels of riding: it is the muscle-tensing and stiffness that create a resistance to the horse's movement and bounce the rider up with each beat. Unfortunately for the horse, this means that the rider will come down again with a bit of a bang in the saddle—and on the horse's back. Not good.

If you are loose but controlled, stretching up from the waist and dropping down from the waist, your seat can move up and down, and dip and rise, with the horse's movements, without ever leaving the saddle. (I have sometimes demonstrated the latter point in sitting trot and canter to clients who have said it is impossible by sitting on a £20 note as I ride and offering to give it to them if it escapes. I have never lost my money yet.)

Loose, mobile hip joints should act like hinges rather than fixed joints, as well, when you are properly relaxed. Your upper body is the top, straight up, part of the hinge and your thighs the lower, moveable part, that go up and down with the movements of your horse's back.

Knowing which foot is doing what, where and when is a priceless ability to have because you will always know just when to give your signal for a transition to a different gait—that is, when the hind foot starting the movement or gait you want is lifting, so that it is free to act. Obviously, if a foot is rooted to the ground under weight in its stance phase, it cannot initiate a gait or movement. Once you know which hind foot is doing what, you are ready to precisely time your different gait aid, avoiding the old enemy, confusion, and possibly anxiety, for your horse because he can respond immediately.

ES has also shown that having control of the horse's legs goes a long way towards keeping him calm. Horses control their direction with their forehand and where the forefeet go the hind usually follow. So, if a horse is becoming excited or upset and starts pawing with, say, his left fore, give a vibrating cue, perhaps with a contact of about 4 on the ES scale, with your left rein only, as a stop request and, of course, continue it till he does stop. If he doesn't stop in a very few seconds, stepping him back, ideally without a break and maybe accompanied by whip-taps on his chest or the front of his left shoulder, both of which you can apply from the saddle, has a salutary and effective result.

You can quickly jump off and step him back in-hand, as you have both done in ES foundation training, but this involves a possible reward for the horse (you getting off), so if you can do this from the saddle it may be better. If he walks off when you don't want him to, try to note which foreleg moves first and give a vibrating rein cue, again, on that side, which should stop him, and then step him back to where he was.

To trot rising, the upper body is brought forward *from the hip joints, not the waist*, so that the shoulders are above the knees and the back remains flat. Think 'hinges'. This position enables the rider to 'rise' by slightly leaving the saddle and tilting the seat-bones forward, then letting gravity bring them gently back down into the saddle. So, rather than going 'up, down, up, down' the rider goes 'tilt, sit, tilt, sit', which, to an onlooker looks little different from a sitting trot except for the forward body position.

This helps the rider's balance and, therefore, the horse's. Also, the lack of a 'thrusting' up-and-down seat movement with an upright body, which can disturb the horse's action and pressure the back every time the rider rises and sits, makes it easier for the horse to move smoothly, in balance and with energy. The rider can take some weight down the inside of the thighs and knees, helping to reduce the rhythmic pressure on the back under the stirrup bars, which can be considerable in a conventional rising trot and must trouble the horse somewhat.

The sitting trot is much easier, too, for horse and rider than in conventional riding because of the controlled looseness of the seat and legs. You don't get bounced up and down due to stiffness, or lack of 'moulding', and can often sit to the bounciest of horses. You aren't actually going up and down, you see: it is your seat that is dipping and levelling up without your having to do anything about it but sit up and stay 'looze'.

Canter the classical way is easier for the horse than the conventional version because of the more rational (in my view) seat and body position. Instead of sitting with the seat-bones straight across the saddle, the inside seat-bone (according to which leg the horse is on regardless of its being a straight line or a curve) is held slightly forward all the time, with the inside shoulder above it. This is the horse's confirmation to stay on that leg and in canter till the rider returns the inside seat-bone and shoulder back level with the outside ones, and it is more comfortable for him, anyway. At that point, the horse will naturally come down to trot without any bit aid being needed, partly, at least, because it is uncomfortable for him to canter with the seat-bones straight across the saddle, out of sync with his back.

As mentioned earlier, during the canter stride, the back is slightly forward on the side of the leading leg for most of the stride so this technique in cantering, and in coming back down to trot, makes sense, and is comfortable, understandable and encouraging for the horse. If you want right canter from trot, for instance, you have to give your canter aid with your left leg (an inward squeeze with your upper left calf) just as the left hind, which initiates the first right canter stride, is leaving the ground. If you have been in (classical) rising trot, sit down and upright into sitting trot, put your right seat-bone (and shoulder) forward a little to warn your horse, give the aid with your left leg as his left hind is leaving the ground in trot during the moment of suspension and off he'll, almost certainly, go! If you want to give a gentle aid/signal with your right rein as well, of course that's fine but not always necessary, depending on the horse.

When performing flying changes, too, this can be done on a horse who knows your canter aids by simply changing over your forward inside seat-bone and shoulder during the moment of suspension after each canter stride, no other aid being needed. On a horse new to this kind of rider, the outside leg aid can be given as the initiating hind leg lifts off but soon dispensed with.

Often, flying changes have to be taught progressively, or 'shaped' in ES terms, but if you and your horse are well-balanced and you always canter with your inside seat-bone/shoulder matching his leading leg (and why wouldn't you?), this way of changing seems very natural and I have known it happen very often at the first attempt, and repeatedly. The moment to do it is as soon as the leading foreleg lifts off the ground, putting the horse into suspension. The horse will 'hop' on his hind foot on that side: this will then make that hind the initiating hind for the new lead. A young friend of mine used to call flying changes at every stride 'skipping canter', which is just what it is.

We often see modern riders, competing at advanced levels, jinking their upper bodies from side to side in flying changes, particularly one-time changes, and applying firm spur aids at the same time. Frequently, too, the rider will give a firm bit aid (double bridle notwithstanding) on the side to which he or she wishes to change. None of this is horse-friendly. It can hassle and upset a horse because there could well be some pain involved in his mouth and on his sides, and it unbalances him. If the horse changes, he probably does so despite his rider's antics. Doing it the classical way, your aids are invisible and your horse is happy.

AIDS/CUES/SIGNALS AND THEIR APPLICATION

Probably one of the most confusing and potentially worrying aspects to horses of modern, conventional riding is the often random application of aids by, in my long experience, most riders and handlers. This is surely because, until recent decades, we have not had evidence of how horses really think and learn, but now we know much more about this and are acquiring more evidence all the time through research.

The term 'aid' was coined to refer to the various messages riders and handlers give to horses to aid or 'help' them to understand what we want—presuming that they actually understand and want to do it. This assumes an attitude of cooperation in the horse that has so far not been scientifically proved to exist. I think that horses who know their riders well and are used to certain procedures probably do want to cooperate, but that younger ones learning the cues and responses may well only want to reduce the irritation of the pressures at first and come incidentally to actually enjoy working with a rider or handler when they get used to everything.

In ES, the terms 'cue' and 'signal' are used as they do not imply cooperation but rather that a horse has been trained to perform certain movements, go in a certain way or fulfil a particular role that we require in response to a cue or signal from us. The aid, cue or signal in a trained horse acts as a stimulus that produces the response we want in the horse. The better trained the horse and rider/handler, the more immediately, lightly and correctly the horse will respond. This is, therefore, a significant improvement in equine welfare.

Stimuli, though, can come from everywhere in the environment and we often have no control over them. Think of thunder, fireworks, someone dropping a metal bucket behind a horse, a frightening vehicle suddenly appearing, a bird flying into a horse's face or, not unknown, someone approaching in a Halloween costume shaking a can of coins. The more confirmed in his responses to our aids a horse is in circumstances like these, the safer we'll both be.

It also follows that if a rider or handler gives a horse painful or frightening stimuli, such as whipping him, jabbing him in the mouth with the bit, kicking him hard with spurs or yanking hard on a headcollar leadrope, particularly on some kind of 'controller' headcollar, he or she is adding to the horse's fear, confusion and anxiety, and greatly decreasing his welfare and well-being, not to mention a smooth, accurate response to cues.

There is, as already mentioned, the important element of classical conditioning attached to responses to stimuli. Horses living near an airport invariably become conditioned/used to the very loud noises the aeroplanes make and don't respond to them. With consistent use of our stimuli, whatever we prefer to call them, horses also get used or conditioned to responding to them as if by second nature or habit. It all means the same thing: the more consistent we are with our cues, the more appropriate those cues are in relation to what we are asking, and the better trained or schooled the horse is to respond to them, the calmer, safer, more confident and more pleasurable to be around the horse will be. Aids given randomly, inconsistently so far as feel is concerned or badly timed produce the opposite results.

I used to ride at a large riding centre that had on its equine staff a former circus pony, Merlin, who had worked with clowns. If anyone stood with their back to the front of his face, he would give them a fairly hard shove: he was apparently neither reprimanded nor rewarded for this and never stopped doing it. I understand that if anyone bent down in front of him, he would also push their bottom so they ended up sprawled on the ground although I never saw him do this. It was such an ingrained/firmly conditioned habit with him that everyone just accepted it, with some amusement, of course.

I was riding him in the outdoor manège one day and my partner was standing leaning over the fence watching. I came cantering down the long side towards him and when we were about one stride away from him, he flicked a carrot out in front of Merlin's nose. He stopped dead in one stride, grabbed the carrot and I ended up around his ears so, rather than raising his head and pushing me back he took the easier way out and dropped his head, so off I slid. I wonder if that was another learned or conditioned response or did he just want to get me off his head? He was a great favourite, being always calm, curious, friendly, fun to ride and completely self-possessed; nothing ever fazed him.

HAND, REIN AND BIT AIDS

Having a firmly established classical seat is a massive help in applying all aids to a horse, and the stability it affords a rider is particularly helpful when applying aids to the horse's head and particularly his mouth, which is just as sensitive as ours.

Horses with so-called *hard mouths* are unlikely to have calloused or nerve-damaged bars (the toothless space between the front and back teeth), as is usually believed, but to be suffering from random, incomprehensible bit aids, persistently firm and unrelenting contact as seems to be so often taught today, incorrectly applied negative reinforcement and/or learned helplessness, all of which have been mentioned earlier. He has become desensitised to it all.

On an encouraging note, I have more than once been faced with such a horse in a lesson and, by using the ES groundwork training described earlier, have achieved in minutes results that amazed their riders, showing that their horses' mouths are really not 'hard' but that their training/schooling has, perhaps through no fault of their owners, been inappropriate. So, if you have or know of a horse like this, I strongly recommend the ES groundwork as rehabilitation, no matter how old or experienced the horse is.

Bitting, bits and the fit of bridles and bits, and also bit contact levels, have already been discussed so let's talk about some other important points.

Ways of holding the reins differ. The usual way to hold snaffle reins is for them to pass up from the horse's mouth, between the ring and little fingers, up the palm and out between the thumb and index finger. The most flexible way to hold them is between the thumb and the middle bone of your index finger. This provides quite firm enough a hold for normal circumstances and enables you to keep your elbows in place by the *sides* of your hips/pelvis while varying the length of your reins as required by opening and closing your lower fingers, enabling you normally to keep your elbows where they belong.

Using the whip while holding the reins is something not to be recommended because you cannot avoid giving the horse a feel or unintentional cue in his mouth at the same time unless you completely drop the contact, which may not be suitable for the moment. If you need to apply a whip-tap behind your leg or on the horse's shoulder, on whichever side, put your reins in the hand on the opposite side so you can hold and tap your whip in the free hand with no concern about giving an unwanted mouth aid.

Whips, of course, must not be used for punishment or to cause the horse pain for any reason. They are an extension of the arm and hand, to reach parts of the horse the former cannot reach. The usual areas in which they are used in riding are behind the rider's leg, either immediately behind to emphasise a leg aid or further back, perhaps with the leg, to emphasise a hindquarter-turn aid. They can also be used on the flat part of the shoulder to help turn the forehand, and on the chest to help with rein-back. If the horse is used to ES foundation groundwork, you can gradually use the whip-taps higher and higher up his forelegs until you are tapping the chest to ask him to move back, so if he feels this when you are riding, he will be familiar with it and hopefully will step back.

Riding with one hand, either hand, is, in fact, a useful accomplishment. The left hand is known as the 'bridle hand' because in battle, riders would hold the reins in their left hand and use their sword in the right. Traditions die hard in the horse world, but as most of us don't ride expecting to use swords these days, we can usefully learn to use both hands singly for our reins.

It is often advised, depending on the school of thought you are considering, to keep the hand, let's say the left for now, in its normal position with the thumb uppermost, the left rein still between the little and ring fingers and then the right rein between the middle and index fingers, the spare ends both passing up the palm and held between the index finger and thumb, with the buckle end hanging down directly next to the shoulder, that is, not hanging over the reins themselves. With the hand in its usual thumb-up position, though, the right rein is higher than the left which must give the horse a slightly uneven feel in his mouth, so I prefer to carry my hand flat, the back of my hand facing the sky, to even things out. With a little dexterity and some practice, you'll soon get the hang of it and it's well worth learning.

The classical direct turn aid is a very subtle, valuable and apparently natural aid to horses, who always respond easily to it, I find. You can give a rein aid to turn without applying any more contact, without the need to open your rein or move your arm out, by simply turning your hand and wrist over outwards, so your fingernails face the sky. Dési used to call it 'thumbing a lift' or 'inviting your horse round'. It is a useful aid just to turn a corner or effortlessly maintain lateral flexion on a longer curve. What happens is that the turning-over of your hand and wrist causes the rein to turn over with it, giving a very slight outward feel on the bit ring or cheek on that side—and the horse flexes and turns that way. Keep your elbow at the side of your hip so your forearm doesn't move out which can cause you to lose your contact, and you'll find the horse can feel the slight turn on the rein. It is subtle, gentle and effective.

The indirect turn aid is another most useful but often misunderstood aid. This is simply a pressing *sideways* on your horse's neck, withers or shoulder of the rein opposite to the direction in which you want him to turn. So, if you want to turn right, you simply bring the left rein in sideways and press it on to his neck or withers. This aid works on the principle, mentioned earlier, of horses controlling their direction with their forehands: it almost 'pushes' the forehand over and you can get a much better-balanced turn than using only your direct rein turn.

Your ultimate turn aid, then, will comprise three moves in quick succession: seat-bone/shoulder forward > direct rein (wrist) turn aid > indirect rein (push) turn aid. If you give each of these three complementary aids at half-second intervals, in a clear and relaxed way, your horse will catch on very quickly to the indirect turn and he will soon become classically conditioned to turning this way, if you do it every time and identically.

The ES way of teaching it would be to use the indirect rein aid first followed immediately or accompanied by the direct rein turn: the horse learns that the new, indirect rein aid is followed by an aid he already knows (direct rein turn) so responds easily and learns the new aid quickly.

A good way of emphasising the indirect rein turn is to quickly tap the neck or withers with the tip of your middle finger on that hand, in our example your left to turn right, stopping the taps the instant he starts to turn (good negative reinforcement). Eventually, you'll find that this decreases to a simple, single fingertip touch as, again, your horse becomes classically conditioned to it, to the extent that that is all you need to do to turn. Believe it.

Things to watch for and avoid in turns are leaning your upper body into the turn, losing your contact on the wrist-turn aid, and pulling back instead of pressing sideways for the indirect rein turn aid. An extension of the latter that you need to be aware of is letting your indirect rein aid turn into the *indirect rein of opposition*. This is where the rider is so keen to get a result from the indirect rein or is maybe doubtful that it is going to work (oh, ye of little faith!) that he or she, in desperation, allows the hand holding that rein to cross over above the withers on to the other side of the forehand. This completely changes the feel of the aid from pushing right on the neck to pulling *left* on the bit, so it actually tells the horse to turn the opposite way to that wanted, hence 'opposition'. As the rider's seat-bone and direct rein turn, whether with the wrist or not, are both telling the horse to go right, our old foe confusion reigns.

In all, it is a case of really considering in detail the effects our aids have on the horse and imagining what's happening as we ride. Remember that horses learn best doing one thing at a time. New aspects can be added and, if well taught, the horse will remember his lessons all his life.

LEG AND SEAT AIDS

As you've seen, the acquisition and maintenance of a secure, balanced, classical seat (which is adaptable to activities other than flatwork) is a major asset to any rider. A tweak to enable you to widen your seat across the saddle (that's right) makes your security and aid application even more reliable and effective.

At the top of each thigh bone, we have a small, stubby bone called the trochanter, which slants forward, upward and inward and fits into sockets at the sides of our pelvis, forming our hip joints. When we mount, if we rotate our thighs slightly so that the trochanters incline a little more laterally than their usual forward position, so taking the tops of our thigh bones a little more outwards, we actually widen the area under our torso that is in contact with the saddle. So, keeping your seat and legs relaxed, turn your left thigh towards your horse's left ear *gently* in a *clockwise* direction and your right thigh towards his right ear in an *anti-clockwise* direction. Don't overdo it (and if there's any medical reason why you shouldn't do it at all, don't).

You will find that this drops you down a little more in the saddle, brings your legs more comfortably closer to the saddle, turns your knees a little inward and helps you keep your toes pointing mainly forward. Keep your seat and leg muscles loose, of course, and just sit there and get used to the feel. You should feel more part of your horse, more like a centaur (but with a better attitude to life!), and better able to communicate with your horse and feel his back and leg movements underneath you.

The legs, seat and bodyweight are powerful aids or signals for a horse. The horse is, of course, balanced horizontally and we are balanced vertically. When we sit on a horse, he has on his back a top-heavy weight. When a rider sits still, with his weight/centre of mass as far as possible directly over that of the horse dropping vertically straight down, and the horse is standing still, it is relatively easy for the horse to bear that weight. The problems start when the horse moves off. (As mentioned earlier, a sound, healthy and reasonably fit horse in mid-life should not be asked to carry more than 20 per cent of his own weight, less for those in other categories.)

The first thing the rider has to do, apart from not interfering with the horse's mouth beyond an in-touch contact, is to relax his seat and legs around the horse and allow them to mould and move with the actions of the horse's back, triggered by the movements of his legs. This sounds easy but, as you may have gathered, either from this book or your experience, it isn't. If you start watching other riders with a keen, critical eye, you will spot myriad 'sins' that make life difficult for the horse, such as bouncing up and down when giving leg aids, not relaxing the seat and legs, not moving with the horse, 'knitting' with the reins, riding in a bad posture, rocking in the saddle, fidgeting, hauling the horse's head and neck in by the bit and many others.

To give a walk aid, cue or signal with your legs, a simple 'walk on' request, for instance, *many* riders put the lower half of their leg back unnecessarily, which raises the knee and disturbs the seat and balance, raises the heel and drops the toe, and aims the aid too far back and too high. The upper body is disturbed by this, too, and the seatbones invariably slide back a bit, out of their balanced, central position. Who would have dreamt that something so simple could go so wrong? Try it this way instead:

Make sure you have a light bit contact. Sit up and still, drop down and loose, open your fingers a little on the reins and, *keeping your lower legs straight down*, give your horse an inward squeeze with your upper, inner calves. If he doesn't respond at once, tap twice with your whip right behind your leg. Let your seat move with him from the start. To ask for more energy or a faster speed, give two quick squeezes in the same way; for longer strides, make longer sweeps with your seat without rocking your upper body around—remember, you're in two halves.

To trot rising, bring your upper body forward slightly from the hip joints, as described earlier, and give one inward squeeze with your inner calves, and two for more energy and speed. For longer strides, rise a little more, reverting to tilt-sit as soon as he complies. Sitting trot aids are the same but your body position remains vertical. For longer strides, you can use your seat.

To canter, say left leg leading, feel for his stride in sitting trot, look where you want to go, put your inside (left) seat-bone and shoulder forward a touch (or more if you need extra clarity), feel his stride and give the *inward* aid with your right leg taken back a bit *from the hip* when his right hind hoof is lifting up, or just about to, for its swing phase.

THE 'CLASSICAL BUTTONS'

You may know of the system of places or 'buttons' along the horse's sides which are used to give leg aids for particular movements. This system has worked well for who knows how long once, as in everything, the horse has been schooled systematically to understand what the different sites mean. Between them, they can be used to ask for forward movement, move the forehand sideways, move the whole horse sideways in various lateral movements, move the hindquarters sideways and during slowing down, halting and reining back.

It is an ES principle to keep things as simple as possible for the horse, so the reins and bit only are used to control his forehand (and the horse's speed, asking for slow down, stop and rein-back). The legs ask for forward movement, control his hindquarters and move him sideways in lateral work, but perhaps with the addition of the outside indirect rein. This is clearer to the horse, makes sense and also avoids two of the most confusing things we do to horses in modern, conventional riding—riding them 'up to the bit' and riding 'forward into halt' which both involve giving 'go' and 'stop' requests at the same time, and which no creature can obey.

STRENGTHENING AND LIGHTENING WORK

The exercises and movements you need to perform for strengthening your horse and encouraging him to lighten his forehand a little himself by taking his weight back of his own accord (the safest and most effective way) are transitions between and within gaits ('inter-gait' and 'intra-gait', respectively) and bending work. As we've said, the horse's spine does bend laterally a little at three places along his back and hindquarter sections (thorax, lumbar and sacral, respectively) but not as much as has been believed previously. This is fine: you, or rather your horse, can still give the appearance of 'bending' to follow any curved line if this is an important judging criterion in your discipline (and perhaps you could ask its officials to reconsider their rules in the light of up-to-date knowledge).

The reason transitions work in lightening is that your horse needs to use and develop different muscle groups to speed up, slow down and bear weight, lightening his forehand as he does so as his centre of balance moves slightly during these adjustments. This work gives excellent results if the horse achieves it purely because he is responding correctly to our signals. If he is held in any kind of artificial posture by gadgets in groundwork or by his rider, his body will not be working naturally and muscle strain and injury can easily occur, particularly during slowing down. His mental attitude to his work can also be adversely affected, producing tension, anxiety and 'unwillingness' (defensive behaviour) during training.

During bending work, including anything from a single corner to a full circle or, when appropriate, lateral work, the horse needs to slightly lighten his forehand in order to move his forelegs more easily around a curve, or laterally. 'Going on the forehand' is actually harder work for a horse because his forehand is weighted and it is more difficult for him to move his forelegs, particularly if even slight sideways movement is required.

The regular pushing, 'leaning back' and weight bearing during this work can obviously be hampered considerably by an unbalanced rider who cannot maintain his or her position in motion and so cannot avoid making untoward movements or weight swings. We are a top-heavy weight on our horse's back so if not only our seat but also our upper body is out of position/balance, this compounds the erratic weight pressures felt by the horse. Horses will almost always move towards weight on their backs simply to stay in balance under it and stay on their feet (this will be discussed more in Chapter 10), which is why weight aids are so very effective when appropriately applied. This also is why so much emphasis has been placed in this book on acquiring a proper *classical* seat, with its foundation in the refinements of relaxation and balance to assist benign control—of both rider and horse.

A SUGGESTED OUTLINE PROGRAMME OF WORK

Let's assume a horse who is half fit due to having hacked around and maybe done some short stints in the school. Depending on your local riding facilities, at home or elsewhere, you don't have to have a school, but it would obviously be useful. (You can do an awful lot of good out hacking.) As well as being the right work to train a healthy, in-work horse to go better, it is also suitable for young horses being brought on, those being rehabilitated after a layoff due to injury or illness (possibly

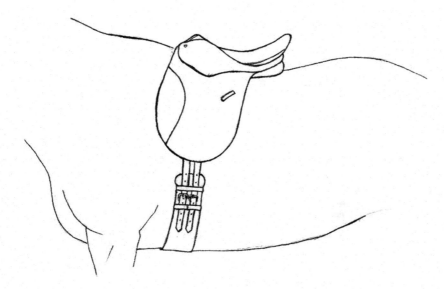

Figure 8.9 A good flatwork, dressage-type saddle. It is fitted to not only clear the withers but also to avoid affecting the tops of the shoulder blades when each foreleg moves back in action. The girth also leaves plenty of room for foreleg movement without digging in behind the elbow. The central seat of the saddle encourages a correct balance in the rider.

in consultation with a physiotherapist or vet) or as a maintenance and age-defying programme for older, semi-retired horses.

The way you keep your horse or horses will have an effect on their work. Those who are fortunate enough to be out a lot with company, particularly on land with different gradients and textures with interesting features and plenty of space, will be fitter and stronger, and possibly more agile mentally and physically, in a natural way than those on a restricted turn-out régime. So you can work them slightly less often than mainly stabled horses from a health and fitness point of view. However, that lifestyle does not keep their weight-bearing muscles strong and fit, so it is not a full substitute for a horse whose work involves carrying weight.

Generally, you need to work a horse on five days a week with two days off, and not together (so three consecutive days and two consecutive days), if they are to make progress in both training and fitness (and with turnout on most days, ideally every day, especially the days they don't work). The exceptions are, first, older horses who, I find, are better worked, appropriately, on alternate days to keep them both gently physically tasked and rested, and turned out as well every day or at least led out in hand twice a day to exercise and graze particularly on the days they do not work, and, second, youngsters who are better worked on about three days a week, spaced out, and given plenty of time otherwise to socialise, play and allow their brains to process their learning. Any creature learns best with regular breaks to allow the brain to create new neural pathways. Sleep, rest and recreation are crucial to understanding and memory retention, for our horses and ourselves.

A POSSIBLE BASIC PROGRAMME FOR RIDDEN WORK

So, we'll assume that your horse is sound and healthy, of course, half-fit, responding promptly to light aids/cues for go forward, slow down/stop/step back, turn, walk, trot and canter, not least due to his ES groundwork in hand. Should any problems arise and any of the horse's responses deteriorate, go back to one stage earlier in the groundwork than the stage dealing with that response and revise forward from there. Also ensure that your aids are light, correctly timed and that you are using negative reinforcement correctly. If you have been taught conventional, modern riding, it might take you some time to get used to a much lighter bit contact and to riding generally on a passive but in-touch contact rather than exerting actual pressure all or most of the time.

Remember your classical seat: keep your elbows back at the *sides* of your hips as their default position; stretch up from the waist and drop down from the waist; neck back and chin in a little; shoulders rolled up, back and down; bottom tucked under a little if you are hollow backed; straight line from the top of your head, through the crucial joints, down through your ankle joint. Before long, this will become second nature and you will return to it like a magnet.

Finally, the different feel of a horse moving with controlled freedom under you rather than being constrained and, shall we say, commanded during his work rather than partnered, may also feel strange, but you will find that you both get much more pleasure and satisfaction from this more humane, effective way of going, calm but enthusiastic, and achieve better results in all your work. You might be surprised at what a lot your horse has to say to you during your sessions together.

WHAT TO DO

When you enter your school, don't make automatically for the outside track. For some reason, horses prefer working here than anywhere else and I think it is because they psychologically come to 'lean' on the fence or wall and find it supportive. However, using the inside track and the interior of the school teaches them to control their own balance better: you might notice a difference of feel in the way they go next to the fence and away from it. It definitely improves their balance to not use the outside track too much.

The Warm-Up

Depending on your horse, it is usually best to begin your work in a long-rein walk with a light contact. If the weather is cold and your horse is clipped (an exercise sheet is recommended), you may wish to go into trot a little sooner than usual to warm him up. If your horse is one of those who believes you should 'start as you mean to go on', in other words 'if we start in a relaxed, long rein walk I'll do that all the time', you might find it better to take a light contact of, say, 2.5 on the ES scale and think more in work mode than warm-up mode, walking in 'working walk' for a start. I know there is officially no such thing, but it is a very useful concept which all horses and riders seem to understand.

After a few changes of direction, and lengthening and shortening of stride, move up to medium walk by increasing the 'sweep' of your seat movement (remember to let your seat move with the dips and rises of your horse's back), giving two squeezes with your inside calves if he doesn't respond to your seat, and then on into working (classical, rising) trot. A relaxed, casual working canter should see your horse warmed and loosened up nicely, all the time keeping your inside seat-bone and shoulder forward, remember, and your light but in-touch contact, changing direction fairly often, and still using the whole school.

Most schooling sessions should be at most 40 minutes long, with several breaks between efforts and definitely after a successful attempt at a particular movement or way of going. A warm-up, therefore, should be completed within ten minutes, but much depends on the weather and your horse.

We are concentrating on gradual strengthening and lightening work just now, leading to bringing your horse on to the bit, which means transitions and bending depending on your horse's level of schooling. You will, of necessity, have done a few 20m circles during your warm-up, so now try working walk to gradually do smaller circles, concentrating on your horse maintaining his speed and rhythm.

If you feel that, on a 20m circle, he has kept his balance and forward impetus and tempo/speed without frequent cueing from you and is 'bent' along the curve of the circle so that you can just see the corner of his inside eye and nostril, gradually spiral him down to a 19m circle and gauge his response. Then go large on a long rein for a minute or so in both directions.

If his head tilts to the outside of the circle and you are sure your contact is not at fault, it could be a sign that he is having a problem balancing on the curves so full circles are maybe a bit much for him just now. Try lifting your inside hand a little to see if that helps, but if he feels to be struggling, do half or quarter circles/curves instead. However, if you don't try the smaller circles, you won't know if he can do them; just don't push things if he can't—yet.

Next, you could revert to 20m circles in working trot and check his balance, rhythm and tempo, come down to 19m and do the same, then have another short break on a long rein. Depending on his fitness, and performance of the walk and trot circles, try a working canter and don't let the tempo get away with you as this will adversely affect his balance. You could find that he leans in with his head and neck out of the circle. This is a clear sign for you to slow down and/or that the circle is too small for him at present.

Leaning in on a circle or curve in any gait can be due to too fast a tempo, a rider who herself is leaning in (putting more weight on the inside seat-bone) instead of keeping her upper body upright, too small a circle, weakness and incoordination in the horse or poor control of his balance by the horse. Simply slow down, go a little larger, correct your own balance, make sure your contact is in touch at about 3 on the ES scale, and try another 20m circle after a short break. Be prepared to weight your *outside* seat-bone to help balance the horse.

It also helps if you look about one-third of the way ahead round your circle. Always be sure that your own position is correct and, therefore, helping your horse rather than hindering him. If these corrections don't work, step back in your programme and give him more time, don't canter, try working trot and if he can't manage that, be content with shallow corners, loops and serpentines. Remember, this can be hard work for the horse. If he's sound and you ride properly, you will both succeed eventually. Don't force things, but keep asking and trying.

Another Nuno Oliveira story: A keen rider came to his yard for regular tuition. On his first lesson, the maestro watched him without saying anything, as was his way. Then he called him in and asked him if he could ride a half-pass. 'Yes, maestro' came the reply. Then Oliveira asked him if he could perform a shoulder-in. Again, the reply was 'Oh, yes, maestro'. 'And can you ride a perfect 20m circle?' was the next question, to which the reply was: 'Of course, maestro'. Oliveira replied: 'Then you are a very fortunate young man. I have been trying all my life and I haven't managed it yet'. So don't worry if you and your horse have problems!

PROGRESSION AND PROBLEMS

Don't be tempted to persist with smaller circles or deeper corners than your horse is ready for. If he can do, say, 15m curves and circles, but not 14m, don't push it. He is simply not ready so go back out to 15m, carry on with his existing work and try again in a few days' time or next week. Remember the 6m limit. Or maybe he can manage a particular curve or circle in trot but not canter, or walk but not trot. It will come in time if you keep yourself impeccably balanced and positioned, keep a little weight down your outside leg, maybe raise your inside hand a little. This goes for any work on curves.

Once your straightforward work is coming together well, you can begin doing more transitions. You need to have been walking in working and medium gaits, trotting in working and medium trot, and cantering the same, all in short stints, calm, relaxed and praising him quietly, rubbing his withers as a reinforcement/reward, and giving lots of short breaks in a free rein walk. Just standing still on a free rein and keeping your body and hands *still* on the buckle is a great reward for your horse. This free rein part is important for mental and physical respite for the horse, and for building trust.

Correct curves, loops, deepening corners and serpentines are fine instead of circles if your horse cannot yet do a complete circle of any size. It will come. Try your increased transitions on straight lines at first, then large curves and so on. Frequent transitions, every few strides, on a circle is hard work and must not be pushed too much.

Transitions within gaits, from shortened strides to lengthened strides and all points in between, are as good for strengthening and lightening as transitions between gaits, but once your horse is fit, strong and very well balanced, the most effective transitions for overall muscle development are those between trot and canter and back repeatedly for half a dozen transitions, then walk on a long or free rein for a few minutes, or just stand, then try on the other rein. A horse will be quite advanced in his fitness and strength by the time he is ready for this, and this will really bring him on so far as being on the bit and going in self-carriage on the weight of the rein are concerned. This work is specifically for strengthening and lightening.

If you seem to hit a block in your work, don't worry. Perhaps you have gone a little too quickly or perhaps your horse needs more different work, free schooling, playing, time off, hacking, imaginative pole work, jumping, different environments for working and so on. Always be ready to give him several days off at a time with plenty of rest-and-relaxation time—remember the Three Fs, friends, freedom and foraging, which should be a major part of any horse's life.

After this, go back a stage in your work and don't actually *work* on several days a week. Hacking and turning out over varied terrain, if available, is great for strengthening and interesting a horse. They don't lose fitness over a fortnight or so off and come back mentally and physically refreshed.

'HAVE YOU GOT A NEW HORSE?'

You will find that with correct work as described earlier, your horse will start naturally to carry his own weight a bit further back, his head and neck will start to stretch and round up and out and he will carry his head, again naturally which is most important, just in front of the vertical of his own accord and because his 'new body' makes it comfortable and inviting to do so.

He will also be very accepting of your aids and, although he may not know it, of your patient and liberal attitude to his work. In the long run, this will make him a better performer—whether or not you compete—healthier, happier (that unproven word again) and a real partner in your equestrian endeavours. Being on the bit will be a normal and natural state for him and a reward and confirmation for you that you have done right by your horse—and, if you wish, are well on the road to more advanced work, which is discussed next.

IN A NUTSHELL

The expression 'on the bit' describes a horse who is going in horizontal self-balance, responding to light aids, maintaining his own speed, rhythm and posture. This description applies to both classical and ES riding.

The slightly rounded or 'flexed' position of 'on the bit' is a result of correct training and work, not of the horse being placed or fixed by the rider to enable him to work properly, which is not how equine biomechanics work, as we have discussed. The horse flexes at the poll joint and is comfortable with his bit and, when dismounted, does not champ at it, put his tongue over it or try to get rid of it. The author's preferred bits, in general, are the eggbutt lozenge bit with shaped cannons and the half-moon (not arched) pelham. The significant evils of tight nosebands are explained and the two or three fingers' width measurement between the band and the front of the horse's face explained and recommended, along with the ISES taper gauge.

Correct light levels of bit contact have been described, and how to judge them using ES and traditional criteria. Qualities of contact are also discussed, those to aim for being 'light, clear and definite'.

Riding 'from back to front' and 'from front to back' are compared, and the feel of a horse going on the bit is described as rather like driving a powerful sports car or speed boat. We have also discussed the three main states of balance of a horse—on the forehand, horizontal balance and 'advanced balance' when the horse carries more

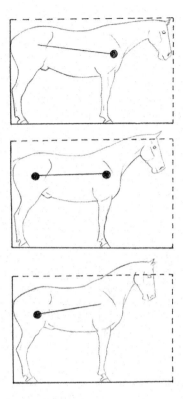

Figure 8.10 Imagining the horse's balance. From top to bottom, we have the forward
balance of a novice or unschooled horse with about two-thirds of his
weight on his forehand, in the middle is the horizontal balance of a
horse who has been in correct work for several months and can go on
the bit with a more even balance, and below is the rearward balance of
an advanced horse, schooled to go with his weight back a little on to his
hindquarters, lightened and lifted slightly in front and probably able to
go in self-carriage on the weight of the rein.

weight on his hindquarters, and how balance and action can be affected by a poorly
fitted saddle.

The physical work has been discussed which strengthens and lightens a horse,
enabling him to work on the bit, and the types of flexion and bend are described and
explained. The author's training with Dési Lorent, a long-term student and friend of
legendary classicist Nuno Oliveira, is described in detail, as being descriptive of how
to attain a true classical seat. How to sit to the various gaits is detailed, plus an easy
way of doing correct flying changes.

The vital topic of aids/cues/signals and their application is treated in detail, what
not to do as well as how to do it effectively and kindly. A programme of work for

strengthening and lightening a horse is given, using transitions and bending work which lighten the forehand through strengthening the right muscle groups. A sample warm-up routine and training sessions are suggested. Solutions to common training problems are given, the causes often being the rider's posture or the horse's needing more time.

Correct work of the type detailed makes a great difference to a horse's body and mind, and he will benefit greatly from it.

9 Adding Finesse—Self-Carriage on the Weight of the Rein

In times gone by, most people wanting to learn to ride well would expect a good riding school to have a couple of schoolmaster horses, or more, on which beginner or novice riders could learn the essential basics, and other horses for him or her to progress to as one's skills improved. In all the best riding academies in the world, novice riders are still taught on accomplished horses with a tolerant temperament and young horses by the best riders, also with a tolerant and understanding temperament. In many riding schools and centres, this is now not so common, but it is well worth seeking out a school with schoolmasters who, with the assistance of a human trainer, can teach beginner and novice riders that foundation which will enable them to progress to more challenging mounts.

Schoolmasters may be used for both learning on the lunge and for general work. The value of a good lunge horse with even gaits, easy to sit to, and used to the weird and not-so-wonderful ways of beginner passengers cannot be overestimated. That said, the opinion has been expressed by some researchers and those involved in equine science that schoolmasters may actually be in a state of learned helplessness, which is why they put up with the uncomfortable and confusing process of being ridden by beginners. In good schools, most only give one lunge lesson a day so there is usually more than one such horse available. Most schools also seem to adopt the policy of giving school horses significant time off to 'freshen them up' before a period of schooling by a member of staff to restore their responses.

A major fault with beginner and novice riders, and sadly some very experienced ones, too, is to use the reins as a comfort blanket, to keep themselves on board and to 'get through' to the horse when they are having difficulties, to the great detriment of the horse's mouth and peace of mind. That is the reason most novice or unbalanced riders on the lunge are not given reins to hold, only a neck-strap until they are more stable and understand more. Schoolmasters are invaluable in helping any rider acquire that essential and priceless possession—an independent seat, something fewer and fewer riders seem to have these days.

Having your own horse may not be the best way of learning or progressing, at any level. A lesson on a good schoolmaster occasionally can give you the feel of what you should be aiming for with your own horse, not least the facility to ride at a higher level, so that the blind is not leading the blind, as it were.

THE ULTIMATE GOAL

The concept of riding a horse in his own self-carriage on the weight of the rein only, in other words no bit contact involving active pressure from the rider, is unfamiliar to

DOI: 10.1201/9781003121190-9

many riders today, but it is by no means new and no one should think that it would be unattainable with their own horse. Without checking back through Xenophon's book, I cannot say for certain whether the ancient Greeks aimed for it, but I doubt it as they had no saddles or stirrups and used the reins and bit to stay on board when necessary. Xenophon was one of the first to propound humane treatment of horses, though, back in what we now call the classical era, whereas in the eighteenth century and before a humane attitude to animals seems to have been mostly lacking (apart from the Duke of Newcastle and François Robichon de la Guérinière), partly because it was believed that they had no physical or emotional feelings, even though the opposite must have been blatantly obvious to any reasonably sensitive person involved with animals.

In ES, self-carriage is stipulated from very early days. Clearly, it is more an all-encompassing term meaning that horses are trained to go in what I call self-balance, not relying on being 'held up' or manhandled around by their rider which, in practice, is a major interference and worry to horses and the antithesis of classical riding and Equitation Science. No wonder there are so many horses with behavioural problems and also hindquarter and hind leg unsoundnesses: they certainly seem to me to have increased along with the spread of the modern way of riding and the demise of many good local riding schools.

Allowing a horse to trust and develop his own balance from Day 1 under saddle, and his being enabled to do so by a correctly balanced rider who assists the horse's balance with his or her own weight distribution, is a great reassurance to a horse. In turn, the rider, knowing that the horse has fulfilled his meticulous ES groundwork programme and/or has been brought on kindly and thoroughly in the classical school, knows that his carefully applied signals are understood and that excellent progress is very likely to be made without too many of the problems that arise from inadequate preparation.

Unforgettable

The first time you ride a horse capable of going in self-carriage on the weight of the rein in its classical understanding, which is voluntarily back on his hindquarters, totally light in hand and instantly responsive, including to weight and position aids, stays with you for the rest of your life, particularly if you have been riding others who had not been brought to that level of working. I have described the feel of a horse being on the bit as like driving a speedboat or sports car, with rear-end impetus under your seat and lightness and responsiveness in your hands, plus light brakes, of course, and it is, but classical self-carriage on the weight of the rein is a whole higher echelon.

A horse on the bit is largely still in horizontal balance, but you will get flashes of the start of further lift in front and lowering behind not to mention an increasing surge of strength from the hindquarters. It can be very tempting to play on and encourage this. Resist! Be patient. Do not increase your contact or raise your hands, do not push your horse further on to the bit and do not pull his weight back more on to his hindquarters. That way, injuries, distress and anxiety lie. It is the way in which many riders, at all levels, obtain a false weight distribution that can be injurious to the horse and without the true performance results or the sublime ride and feel of a carefully, patiently and correctly trained horse.

Figure 9.1 Secret just learning collected trot in self-carriage on the weight of the rein.

Keep your in-touch contact, bearing in mind that as your horse's carriage changes and his head and neck are carried more arched upwards as well as outwards, with the nasal planum more naturally approaching the vertical, you may need to slightly shorten your reins so that you do not '*abandon your 'orse!*'. The fact that the horse feels that he needs or wants and is able to adjust his—and your—weight in this way is testament to your good training and riding. By all means positively reinforce it by saying, in your usual quiet, praising tone *as* he does it: 'good boy!', but don't actively ask for it—because you don't need to. Keep it simple. You are clearly on the right track and it *will* come naturally if you carry on with the strengthening and lightening work up to the level he can manage—and that level will rise as a matter of course as the result of your training.

Problems, Problems!

If you meet a block and don't seem to be getting anywhere, some weeks off training, a complete rest ideally in comfortable conditions in the field, or a change to only hacking often does the trick. Consult your trainer (probably a good idea to avoid social media gurus!) or maybe think about a different trainer if things are feeling stale, but put your horse's needs, as shown by his reactions to his work, first. Always be prepared to blame yourself first. Let your horse dictate your schedule.

If you find one particular movement or exercise is a problem, say 10m circles, go back a stage a bit further than the point at which the problem started. If you first noticed difficulties at the 10m level, revert to 11m or 12m without any psychological or physical pressure. Take it slowly and easily and not too often. If any problem is persistent, consider consulting a vet or physical therapist to check the horse's comfort and soundness. Have his saddle, shoes, teeth and feet checked, too. Remember that larger horses can have problems in athletic training sooner than smaller ones. So, whatever your problem, go back a stage and work on to the new level, and if it occurs again, seek expert advice.

THINK IT AND IT WILL HAPPEN!

The feeling you have on a horse who has moved on to being in self-carriage on the weight of the rein is that you have at your disposal the most powerful and sensitive living machine imaginable. The horse is ultimately in charge of himself now, but you are in the driving—sorry, *riding* seat. True, other animals are ridden and are equally sensitive, but none can be trained or perform in the way a horse can because of his unique assemblage of physical and mental qualities.

You may have heard that you only have to think about a move on such a horse and it happens. There may well be, as I have read, a physical phenomenon involving an emotional, neural and physical chain of events that occurs in a fraction of a second, which means that just thinking of a movement or aid results in the rider's muscles very faintly acting (unconsciously on the part of the rider), and in a highly trained horse responding accordingly. Whatever the case, riding such a horse really is like magic.

Figure 9.2 This lovely photograph is of Sylvia Loch and her Lusitano stallion, Prazer, illustrating a horse going well on the bit. Riding with one hand is a useful skill to acquire, as detailed in this book.

Physically, you begin to feel that your horse's back, your base of operations, is widening slightly. This is because of his more advanced development and way of holding himself, and of his body compressing naturally and slightly as the weight shifts back, musculature developing and his stance and action becoming more stable and 'purposeful'. You feel that you are being *very slightly* repositioned back a touch in the saddle because of the horse's increased rearward weight carriage, lowered quarters and raised forehand. It's not enough to disturb you and you do get used to it, recognising it as confirmation that you've 'got it', as when you were learning the classical seat which is essential to ride such a horse well. As a good rider, you will keep checking and correcting your position, anyway. Your weight aids and, ultimately, every slight shift of your seat, thigh and calf positions prompt the horse to respond to them as he thinks fit. You may start to sense from the horse that he is saying: 'This is me and this is what I do'. And he might ask you: 'And what do you do?' This confident attitude in your horse, of course, adds to your responsibility because you have to live up to what you have created together.

Your reins, whether of the proverbial silk or, more likely, leather, whether single snaffle reins or double-bridle/pelham reins of bridoon and curb, simply lie in your hands and you are able, much of the time, to ride in *descent de mains*, a French classical term meaning descent of the hands. This can occur when the rider has cued the horse to perform a certain movement or gait, the horse responds and the hands

Figure 9.3 Anne Wilson on her very experienced mare, Lucy, in rein-back. Anne has lightened her seat to free Lucy's back and hindquarters for the rein-back.

are relaxed and lowered as the rider sits there, perfectly balanced in that controlled relaxation, allowing his or her seat to move with and absorb the horse's movements but appearing as still as a statue, enjoying the sensation—having correctly applied negative and positive reinforcement, knowingly or otherwise! By this point in his work, the horse will know to keep performing the movement until his rider indicates relaxation or a different movement, hence the dropped hands.

'THAT'S USEFUL TO KNOW!'

When horses absorb the training they receive and begin to feel stronger and better balanced, it is surprising how sometimes they use their newfound abilities to further their own ends!

I used to be privileged to ride and school a Fell Pony mare who could be a 'right little Madam', very intelligent and quick-witted, like many native ponies. We had two particular rides which both started at the end of one track, one going straight ahead on to a longer ride and one turning left for a shorter one. We were trotting gaily along the track and my intention was to go straight ahead, which she could tell. In fact, the second choice never entered my head—but it entered hers! As we drew level with the left-hand track, she suddenly slammed on the brakes, half-reared, sat on her hindquarters and then leapt forward and left into a fast canter before I realised what she was thinking. I just about managed not to fly off over her right shoulder.

She wasn't the only one who has learnt to put her classical training to good use in getting her own way or achieving some aim or other, including in the field. Horses—and ponies—presumably realise that they are capable physically of more than they were previously and take full advantage of their newfound strength and agility.

A habit of the previously mentioned Fell Pony, also, when she didn't want to do something was to raise herself up into a sort of natural levade and just sit there for quite some seconds before resuming all-fours. Usually, after she had done that, she would do whatever it was she hadn't wanted to do. Anthropomorphically speaking, it seemed that she just wanted to make it clear that she would do whatever it was you were asking, but under protest.

THE GAITS AND THEIR BASIC VARIATIONS

SOME POINTS TO CONSIDER

Gaits have been discussed earlier, but not their main variations. The term 'ordinary' to define a gait was abandoned some decades ago, maybe because of its imprecision. Now, gaits are termed more precisely as working, medium, collected and extended. Note that extension comes after collection—it has been described as collection with longer strides and a horse needs to be well-established in collection before being asked to lengthen, *gradually*, into extension without falling on to his forehand.

The whole point of schooling and training, down the centuries, has been certainly to train a horse to fulfil a role but also to make his gaits *enhanced versions of his natural ones achieved under weight*. Letting a horse be the very best version of himself should surely still be the purpose of schooling and training but instead, today we often see so-called 'spectacular' action which, to true horse lovers, is upsetting, ugly and must surely be potentially damaging to mind and body—my opinion.

THE WALK

A four-beat, lateral gait in which the footfall sequence is left fore, right hind, right fore, left hind, the walk is officially described in formal dressage terms as being medium, collected or extended. I add a 'working walk' to that because it can be such a useful gait for developing carriage and relaxing a horse. Done well, an energetic, 'going somewhere' walk is excellent for initial fitness training, including hill work, without the concussion to limbs that can come with faster gaits.

It is also a very useful gait to give a horse time to think about the pressures, aids or signals applied to his body. Of course, giving an aid at the precise moment when the horse can comply *is* important, but so often I have found that putting yourself into the correct position, giving the aid(s) and maintaining this state without hassling the horse gives him time to think and he will usually make an attempt without getting his feet in a tangle, and get it right. This polite waiting for a few seconds while maintaining our 'ask' by body position and continued aid application can be transferred to the faster gaits with great advantage, and it seems to help horses, knowing that they are not being nagged. It follows the ES principle of maintaining your aid till the horse responds, then releasing it (negative reinforcement) which is a reward in itself, but ideally also giving primary and secondary positive reinforcement, as described, such as 'good boy' plus a rub of his withers.

Walk is also invaluable for the rider to get used to thinking of his or her body in two halves softly connected at the waist. In this gait, we can sit stretched up from the waist and drop down loosely from the waist, and maintaining this concept in our minds as we ride seems automatically to keep our upper body still, all the movement being absorbed by our relaxed seat and legs. We can learn to give better aids than perhaps we have used in the past—an inward squeeze with the inner calf for go and two rapid ones for more energy, a longer sweep with our seat *without* rocking our upper body for a longer stride, and a light bit aid to steady or slow the horse, or halt, increasing the pressure slightly, if necessary, and then vibrating quickly at that pressure, which totally makes pulling redundant. Practising our turn aids—direct rein turn with classical wrist-turn for subtlety, and indirect rein turn, for a better balanced turn.

GOING WITH THE SWING

In walk, horses naturally swing their heads slightly from left to right and up and down, but the usual way of 'going with' this movement is quite disturbing to a rider's position and seat because most riders 'give' rhythmically by putting the hands, and unavoidably their elbows, forward and back with the rhythm of the gait. Keeping the upper arms *vertical* and the elbows back in their default place at the *sides* of the hips helps to keep the rider's torso still and prevent it rocking forward and back with the rhythm, giving an unwanted weight aid to the horse. So instead of putting your hands and elbows forward, open and close your fingers alternately to allow for the head movement.

Figure 9.4 Here, Anne is starting to resume her normal position, her legs dropping down ready to give the walk-on aid so that Lucy will move forward smoothly after stepping back.

Figure 9.5 In this photograph, Anne was asking Secret for collection—Secret briefly gave her the feeling she was seeking, so Anne then sent her forward at a brisker trot.

Your horse will naturally, if on a loose or long rein, swing his head left as the left fore lands and vice versa. On a long rein, you will still have a gentle but not wishy-washy contact. As the head swings left, even just a little, your left rein will feel slightly slacker and your right slightly tighter. This means, of course, that your horse is feeling bit pressure alternately on the 'tightened' side of his mouth instead of an even, consistent contact. You can certainly, when learning the finger technique, slightly open and close the fingers of both hands at the same time, but if you can refine the technique, as follows, you will keep an *even*, light pressure on the bit despite the swing, and avoid giving your horse alternate, distracting and possibly confusing increases and decreases in bit pressure.

So, as his head swings a little to the left and your rein feels a bit looser, close up your left fingers to take up the slack—*and at the same moment, open your right fingers to release that rein a little because it will feel tighter as the head swings left.* Then his head will swing right as his right fore lands, so now your *right* rein will feel loose and your *left* tighter, so you alternate by opening your left fingers and closing your right ones, both at the same moment. Therefore, you will be going along opening and closing your fingers slightly and alternately to accommodate the changing lengths in your reins, and pressures in your horse's mouth, as his head swings slightly to left and right in long-rein walk. The end result is that you and he will feel an even bit contact throughout the gait, stabilising his gait and removing any chance of irritating or confusing your horse.

You can use this technique in medium, collected and extended walk, too, because although the swing is not so noticeable because of your contact (which is even lighter in long-rein walk), the tendency is still there for him to naturally want to swing his head in accordance with his gait to balance himself. You will find he gives you much more precise and 'together' gaits if you use this technique. And when you do give an aid, he will know it's an aid.

This does take a bit of thinking about, and practice, but it is a subtle refinement that increases a horse's trust in the bit, which consequently retains a stable feel in his mouth without being rigid, of course. Anything that helps our horses is important to master.

THE TROT

Trot has working, medium, collected and extended versions. It is the trot that can be most problematic as this is the gait in which some trainers and riders confuse 'forward' with 'fast', ending up with horses flattening out, off the bit, flicking up their toes and falling on to the forehand, consequently pushing on to the bit for support. If a horse needs balance support from the bit, he is not ready for whatever he is doing. (It's noticeable that trainers and riders are not so keen to perpetuate this error in canter!) The toe-flicking action of the forelegs sometimes seen can stretch and stress their precious tendons and ligaments. I have never seen a horse going this way who seems to be enjoying it, which says a lot.

To avoid this considerable and very common problem, simply don't think of driving energy in trot. Fix in your mind that 'forward' means 'on the aids' or responding quickly and lightly to your aids, whether they ask for forward, backward or whatever else. When a horse *is* going too fast for his own good or 'faster than he can', you will surely feel it. The stability of his posture and movement could deteriorate into a forward and downward feel in front. The hindquarter impetus of a horse on the bit, pushing your seat onward and upward, will be replaced by a general feeling of unsteadiness, being out of balance and even out of control of his own body, and too much pressure in your hands, from the bit obviously.

If at any point you feel anything like this, immediately apply slowing down aids which means no legs (obviously) but a clear and maybe vibrating bit aid as described earlier, accompanied by whatever vocal signal you use to slow your horse. Keep your head and don't give a sustained pull as he could well lean on this all the more. It may well help to slow your rise, or tilt, which encourages horses to 'come back' to their rider. If you are in sitting trot, move your seat fractionally slower than his back dips, which has the same effect. Another thing you can do is to raise your hands slightly, depending on his head position, as you give your bit and vocal aids, and continue them till he slows down, maybe just to a working trot; then ask for walk and let him calm down. This could have been a frightening experience for him. Soothe him with strokes and kind words.

RIDE AROUND YOUR HANDS

Another rider error which can completely put a horse off his stride is for the rider to allow his or her hands to go up and down with their body, almost anchoring on the horse's mouth, as they rise and sit to the trot. If you use the classical 'rising' trot described earlier, you will hardly be rising and sitting at all, just doing the tilt, sit, tilt, sit movement, with your hands having no need to rise and drop. Horses do not swing their heads around in trot, but keep them still and stable. So, regard your hands as a still, focal point in trot, held in the same place, and think of riding 'around your hands'. This attitude helps greatly to keep a soft, consistent contact in all variations of the trot.

THE CANTER

Canter similarly can be working, medium, collected and extended. The most common problems I think teachers come across with canter are striking off with the correct leg when a horse has a clear preference for the opposite one, and staying in canter with a horse who keeps coming back down to trot. First, the saddle and girth fit and positioning should be checked (discussed earlier) as discomfort with either can be very off-putting. Then it should be considered that the horse could have a problem with his back, also maybe with the hind leg starting the canter and the leading leg finishing the stride, both of which carry all the weight of horse and rider at those points in the stride, so are under particular stress.

If these points are not issues, the cause is probably that the rider does not know about the positioning of the seat for canter (most modern, conventional riders don't), and it could be a combination of any of these. Another, rarer, cause is that, in exasperation, the rider gives the leg aid for canter plus a hard jab in the mouth on the side of the required leading leg, for example, a hard left rein aid for left canter.

Figure 9.6 Here, Secret herself offered some collection, her head and neck arched upwards and she was briefly in self-carriage. Anne did not seek to sustain this for very long as Secret here was in the early stages of this training.

The ideal way would be for the trainer to try first to help the horse, with correct techniques of signalling and timing, to understand the aids. The trainer would then explain the seat positioning to the rider and, if possible, give him or her a lunge lesson on the procedure, and repeat it without the lunge. Off the lunge, the seat positioning should be fully explained to the rider and practised in walk and trot on turns. Then the horse should be asked for canter on the side he favours, with the rider adopting the correct seat position. Next, canter on the 'difficult' leg should be asked for, with the rider positioning the inside seat-bone and shoulder forward in a slightly exaggerated way, and weighted slightly.

As explained previously, the leg aid should be given just as the hind leg starting the stride is lifting up. Then it would be a big help if the rider immediately could give an indirect turn signal on the outside/opposite shoulder or withers when the intended leading leg is lifting. This nearly always, if not the first time, results in a correct strike off, assisted by a very definite forward positioning of the rider's seat and shoulder on the relevant side. It reads more complicated than it is in practice.

For a rider who gives a rather harsh bit aid on the side of the proposed leading leg to 'help' the horse strike off as required, this can hopefully be stopped by the trainer's explaining how painful this is for the horse and that it will not actually be useful. In fact, we can often more successfully enable a horse who is having trouble striking off as required by letting him turn his head and neck slightly to the opposite side because this is his natural way of functioning (you can see it in horses in the field).

Figure 9.7 Collected canter prior to training the pirouette. Here, Lucy is leaping off the track to make an acute turn—not exactly correct but a very good attempt.

If the rider seems to be somewhat fixated on his or her hands, the teacher should try to direct attention to the seat and leg aids and positioning and, himself or herself, pay considerable attention to correcting the rider when the hands are used, or likely to be used, roughly. If all fails, lunge lessons with a neck-strap but no reins should produce better balance, more confidence, and less attention and reliance on the reins. Riding in a straightforward, side-pull type of bitless bridle would produce less pain for the horse.

ES 'LITTLE' GAITS

In Equitation Science, there are very useful 'little' gaits, which are shorter versions of the standard working gaits but still in rhythm and with your usual in-touch bit contact and light aids when required. They are mainly used with newly backed 'beginner' and novice horses to allow them to find their feet and their balance under a rider in the different gaits—so you can use little walk when the horse is just doing walk work, little trot, which I find particularly useful, and little canter, which some horses find tricky because it is hard to balance under a rider at first, even if the rider's balance is impeccable, so don't shorten too much and keep everything relaxed.

THE REIN-BACK

Rein-back, although not a gait as such, can be misunderstood and not performed well. It is important for manoeuvrability of a horse so deserves working on. It is also useful to relax and reassure keen or anxious horses by stepping them back and forward a few times if in-hand, or performing a calm, gentle and correct rein-back if mounted (no leg pressure, of course, because you don't want him to go forward), a short walk forward, then another rein-back.

Rein-back is not a backward walk but more like a backward trot without the moment of suspension—two-time, with the legs reversing in alternate diagonals. Major problems occur because of the modern practice of applying go and stop aids at the same moment. Horses are pushed up to the bit into halt, then often have to be firmly pulled back by the bit, rather than their responding to a gentle stop/slow/back cue, purely because the rider is applying leg pressure as opposed to maintaining passive contact down the horse's sides. Horses often squirm into halt with splayed hindlegs because of the impossibility of going forward and backward at the same time.

Riders who accept that this isn't working very well and soften their legs to a passive in-touch contact with the horse's sides, and can also realise that the hard, modern contact obviates light aids, do much better. They can give a light bit aid from halt, maybe with a soft vibration and lightening the seat a little, and the horse will understand and go back in response, when the rider must release/stop all aids to correctly reinforce the response. The halt needs to be achieved similarly, with no leg pressure, for the same reason.

To train rein-back, and halt from walk if this, too, is problematic, have a helper on the ground with a schooling whip. Give your bit aid only, accompanied by whatever

Figure 9.8 Prazer maintains his own posture on a soft rein, as Sylvia thanks him by rubbing his withers.

Figure 9.9 An advanced horse, wearing a pelham bridle and going well in canter, self-carriage on a very light rein. His weight is clearly back slightly on his hindquarters due to his correct work, with *no* force from his classical rider. Note his head and neck are arched and stretched up and out, his poll is the highest point of his outline and his head is in front of the vertical.

word the horse knows for going back—such as 'back'—and have your helper gently, quickly and silently tap him on the front of the cannon furthest forward (because that's the one he'll move back first) with the end of the whip. Having done ES groundwork will make this a virtually guaranteed halt and rein-back.

Then, do up to five repetitions with the helpful whip-taps, higher and higher up the legs and finally on the chest. This is so that, if necessary, you can reach forward from the saddle and tap him on the chest to ask him to move backwards, and he'll understand.

SEAT AND WEIGHT AIDS

Ah! The invaluable and essential basis of classical riding. Years ago, I watched time and again a video of a Spanish rider giving a demonstration of schooling and riding without reins on a horse retired from the bullring, in exactly the way we were taught by Dési, with the specific purpose of demonstrating the power of an exemplary classical seat and a horse going in self-carriage on the weight of the rein.

The horse did have single curb reins, but they rested on his withers and looped down nearly to his elbows. Horse and rider entered in Spanish trot, the rider's hands on his thighs, to the centre of the arena, halted from Spanish trot, struck off from halt into a very collected canter and gave a riveting demonstration of all the movements required of a bullfight horse—turns, half-passes, full-passes, flying changes at every stride, pirouettes with a flying change and a pirouette the other way round, canter in reverse, then on into a gallop around the arena before coming down to canter, halting from canter and, from halt, leaving the arena in *passage*.

I have seen similar demonstrations before and since this one, but the horses have often seemed anxious, worried, wide-eyed, sweated up and on pins, with a good deal of domination involved. This rider and horse were different. They were both calm and demonstrated their skills with total confidence, concentration and accuracy as if to say: '*That* is how it's done'. Most noticeable was the faultless upright carriage of the rider and the positioning of his body in the turns and bending work—no leaning into the turns which can actually bring a horse down but, because I knew what to look for, a clear forward placement of his hip and shoulder on the side towards which the horse was turning, more obvious the sharper the turn, plus the occasional use of an indirect turn aid with his hand on the horse's opposite/outside shoulder, just as Dési had taught us—with our reins on a shallower loop, though.

The horse's way of balancing himself was also interesting. Horses, of course, use their head and neck as their 'balancing pole' serving the same function as our arms— deprive them and us of that facility and moving is so much more difficult. This horse, of course, was completely free of any head influence. Sometimes, when travelling forward on a curve, he turned as in a conventional dressage way, bent and looking into his turn, occasionally muzzle leading. At other times on tighter curves, he led with his shoulder, and his head and neck turned slightly the other way to balance himself, the rider helping by *not* leaning in but infinitessimally out. In the pirouettes, he seemed naturally to do them as they should be done—with the *hip* leading, the shoulder just behind and the head and neck bent very slightly in the direction of movement. The rider did not make the conventional mistake of leaning in the direction of the pirouette.

TRYING TO USE 'A LITTLE LESS HAND'

Seat and weight aids are, clearly, very powerful. I doubt that anyone would want to rely entirely on them in an emergency such as a suddenly frightened, panicked horse, or one you did not know well, but if you have your reins as a back-up, and maintain your in-touch contact on horses who cannot yet go in self-carriage on the weight of the rein, it seems to me that most of us should be directing our horses more with our seats and, in Oliveira's words, 'try to use a little less hand'.

It is important to remember to keep the upper body upright, shoulders back and down, chest expanded a little, and the whole torso held in the hopefully now familiar controlled relaxation. This should be at the forefront of our minds until it becomes second nature because every minor slouch affects the spine and the rider's balance, which is made worse by not keeping the elbows back in place at the sides of the hips/ pelvis.

Applying a weight aid is done just by stretching one or both legs, as required, down into the stirrup(s) to bring more weight on to the seat-bone and stirrup bar, and to position the seat-bones where ever they are needed. This is when it feels natural to lean over a little, but don't do it. Stay upright to help your horse balance.

Riding curves is often done in quite a dangerous way. Remembering that we are top heavy weights on our horses' backs, riding on a curve, so useful for lightening

the forehand and getting his weight back, of course, should not entail the rider leaning into the bend as this will over-weight the horse and incline his balance inwards
but not in a good way. Depending on his speed and the tightness of the curve, he
may well have to scrabble around with his feet to stay on them and orient his head
and neck to the outside to counteract our leaning. To help your horse and accomplish
curves more easily, stay upright.

If you watch footage of barrel racing, you will notice the extremely tight turns the
horses have to do at high speed, and how all the best riders stay upright no matter
how steeply their horses tilt their bodies to get round the turns. If their riders leant in
with them, they would surely lose their footing and fall.

Watching horses playing at liberty, they use their heads and necks freely and naturally turn them out on tight turns. If you are involved formally in the discipline of
dressage, it is stipulated that horses must follow the curve with their heads, necks and
bodies, presumably adopted because it looks more attractive and logical to human
eyes. Your horse will find this much easier if you stay upwards, relaxed and controlled. You can put your inside seat-bone forward and weight it a little to start the
curve, but bring it back level with the outside one once *on* the curve, along its radius,
maintaining the curve with the wrist-turn and indirect turn tap or sideways rein push,
as needed. If your horse is drifting in, weight your outside seat-bone.

The exception, of course, is canter when the inside seat-bone needs to remain forward to accord with the leading leg. If you remember the saying: 'Where you look
and where you put your weight, your horse will go', you will find it easy to maintain
the curve in canter. Look about a third of the way around your circle or curve and
equally weight your seat-bones, but if your horse does drift in or out, you can quite
easily counter this by putting more weight on the opposite seat-bone while still keeping the *positioning* correct.

So, if he is drifting out left, weight your right seat-bone, tap his left shoulder, and
vibrate the right rein—and vice versa. This is much more subtle and effective than
just using the bit and direct-turn aids to correct it. You can maintain your curve with
the wrist-turn cue.

FINAL THOUGHT

Riding a horse in self-carriage on the weight of the rein should be every horseman's and
horsewoman's aim because, achieved properly with the horse's musculature and balance properly developed, it helps to maintain the horse's soundness and produces an athletic, agile horse plus ease and joy for the rider, and, I believe, for the horse. Advanced
movements cannot easily or correctly be performed without it, and it cannot be forced.

It is not necessary, of course, to be able to produce High School movements. Simply riding a horse in self-carriage, or even just self-balance, on a light, in-touch contact and going on the bit is excellent, much more enjoyable for horse and rider than
the modern norm and will give the previously mentioned benefits.

IN A NUTSHELL

It is a great advantage to find a riding school that has schoolmaster horses on which a rider can learn to develop an independent seat, no matter how experienced he or she is. Having your own horse may not be the best way to learn or progress. Any healthy, sound horse of reasonable conformation, given the necessary time and correct training, can be developed to the point of going in self-carriage on the weight of the rein. In Equitation Science, self-carriage or self-balance is cultivated from the earliest days of work under saddle. The current fashion for 'holding together' a horse in an enforced posture to look as though he is in self-carriage or at least in collection actually prevents a horse balancing himself and can be responsible for physical and behavioural problems. A horse able to go on the bit is properly prepared to progress to classical self-carriage on the weight of the rein which, when achieved, is a feeling riders never forget.

If a horse experiences problems in training, go back along the training programme (clearly set out in ES) to the level before the problem started, or further, if necessary, to retrain earlier levels which may not have been properly absorbed. It is worthwhile to have the horse checked over physically and to give him some time off or doing different work, more hacking or whatever he enjoys.

The object of training a horse is to enhance his natural gaits, not to develop artificial and possibly rather strained versions of them. The main four gaits have been discussed with information on achieving and developing good ways of going. We have also discussed rein-back, ES 'little' gaits which are very useful for the learning process, the rider's seat and weight aids, riding curves emphasising the importance of maintaining a correct, controlled and relaxed classical seat to help the horse work well, the rider's balance and position being crucial to this.

10 Fast Work and Jumping

It might seem strange that the principles and techniques of classical riding, with its image of 'fancy' work and airs above the ground, could be applicable to fast work and modern show-jumping and eventing, also racing, but that is, indeed, the case. Equitation Science, of course, is applicable to these topics and every other relating to the training, care and management of horses.

Classical and ES principles are both a great help in the faster gaits, up to and including racing, where balance and control are crucial. As with riding on the flat, jumping techniques have changed significantly and, in my view, much for the worse, over the past few decades. I hope the views and experiences I give in this chapter will be of interest and help to readers.

FAST WORK

At what point can we call a horse's gait or gaits 'fast work'? The description is applied to fast cantering and galloping including that in-between gait called a three-quarter speed gallop.

The canter, of course, is a three-beat gait, starting strides (one stride being four footfalls) with a hind leg, then the other hind leg accompanied by its opposite fore and finally the remaining fore—so, starting with left hind, then right hind with left fore, finishing with right fore followed by the moment of suspension. This shows that two legs in canter, the hind leg that starts the stride and the foreleg that finishes it, singly bear all the weight of horse, rider and tack at certain points in every stride—and at speed. If a horse is having difficulty on one canter diagonal, therefore, but not on the other, it is possibly one of the two legs that singly bear weight during the stride that is the source of the problem.

A formula you might remember from physics classes at school goes: force = mass x acceleration, or in non-scientific language, force = weight x speed, therefore the faster a horse is going and the more he and his rider and saddle weigh, the greater the force applied to his feet and legs and, of course, further up his body. In gallop, this is even more of a point because the feet all land singly on the ground, all at some point in the stride bearing all the weight with the even greater force of the increased speed.

Horses do not show lameness anything like as clearly in canter as in trot. Even very mild lameness shows in trot, especially on curved lines and hard surfaces. If a horse is noticeably lame in walk, the lameness is usually more significant.

Extended canter in dressage terms is not called fast work or a fast canter, although clearly it is faster than medium canter, as in ground covered in the same time due to the longer strides. A fast ordinary, casual, hacking canter, sometimes known as a *hand-canter*, is still a three-beat gait or it would not be canter, but horses are quite

DOI: 10.1201/9781003121190-10

capable of extending out and still using a three-beat stride pattern, which is a really enjoyable gait to ride.

It can be moved up to a *hand-gallop* which is fast and just into the four-beat gait but not so fast as a *three-quarter gallop*. It is useful for hacking on suitable tracks or going over any open, safe country with or without fences.

I remember one cool August afternoon in the 60s when I took my horse, Royal, onto our local beach, which is seven miles long and perfect for riding about an hour or more after the tide has gone out. We reached the water's edge and he had his usual splashy paddle, sniffing and licking the surf. Royal was pretty fit and coming up to his prime at that time: as an Anglo-Arab, stamina was in plentiful supply. He came out of the water and set himself off into a hand canter south, well within himself, and kept it up for about five miles, then he suddenly went into a steady gallop: what I can call a hand-gallop. I just left him to it all, wondering how far he would go. He whizzed on for about a mile, after which he came down through his gaits to walk, took a deep breath, shook himself and went back into the sea. He turned north for home and we walked in the shallows for a couple of miles (excellent for the muscles up to a depth just below a horse's knees). Then he came out, set himself off in a hand-canter again until we reached our exit point four miles on, and walked out all the way home on a long rein. That was one of my most memorable rides with him. He often decided our route, and I was happy that he was happy doing so. Royal's local nickname was 'The Horse Who Likes People' because he'd stop and talk to anybody on our ride, not just for titbits. He was gelded late and always behaved more like a stallion than a gelding: when he was turned out with the gang, he would separate the mares up one end of the field and the geldings at the other, and patrol the space in between, keeping the mares for himself!

Gallop is, of course, a fast four-beat gait in which each leg, at some point in the stride, is bearing *all* the weight of horse, rider, saddle and any added weight (or 'lead') to conform to the requirements of the race or discipline involved. The combined requirements of supreme cardiovascular effort and weight-bearing on all four legs singly makes the gallop a gait for which a horse has to be prepared properly and gradually as regards athletic fitness and strength. The sequence for the gallop starting with the left hind, as in canter previously mentioned, is left hind, right hind, left fore, right fore, then suspension. The appearance of the extra beat occurs in the original two-leg middle beat of the canter, in which the two hooves separate in the gallop—in our example, between the right hind and the left fore.

When considering pure, correct gaits, the point at which canter moves on to gallop is determined by the number of beats in a stride. An over-collected canter and riding horses over-flexed in *Rollkür* can result in what is called 'diagonal advanced placement' (DAP) where the hind foot of the 'dual' beat lands before the fore, resulting in four beats as for the gallop, but obviously slower; however, despite the four beats, it is still called canter but an impure gait. This can also happen in trot for the same reasons, the hinds landing shortly before the fores. Any kind of forced riding can result in impure, contorted gaits, including a 'disunited' canter, called a 'round gallop' in some countries. The footfall sequence in this is, for example, left hind, right hind, right fore and left fore: it is most uncomfortable, so the rider can always tell when it has occurred. Domineering and restrictive riding can also trigger defensive conflict behaviours commonly called 'bad behaviour' in the horse, but probably due to distress and confusion.

So, if we are thinking in terms of pure, correct gaits, the point at which canter becomes gallop is when the three beats become four. This point can usually be felt by most experienced riders, and it can certainly be heard if you are on a surface that makes a sound when the hooves hit it.

Where, then, does this leave the useful *three-quarter speed gallop*? Well, it is still a gallop with four beats but not the fastest gallop of which a horse is capable. Its recognition comes as a result of that admittedly sometimes elusive quality in the rider of 'feel', but this can come with experience. The three-quarter gallop occurs, in my experience, when you feel the horse is swinging on and making an effort but, as it were, 'still has something left in the tank' if you press the accelerator. Horses trained in that speed of gallop become used to it and rarely exceed it without being asked, unless possibly galloping 'upsides' (next to others) during training for racing, which is exciting to them and can hype them up to go faster. Control at this point needs to be practised, of course, as 'losing' your horse in a race can land you both in trouble—and not only in a race.

A full, flat-out gallop is usually only seen if a horse is galloping, or 'bolting', in fear or is being pushed or 'driven' in some competitive situation, such as the finish of a race, although some eventers go at a fair lick if needing to make up time. In usual domestic conditions, most horses do not get the chance to perform an extended, flat-out gallop, or riders to experience it. My riding on my local beach decades ago gave me plenty of experience of full gallop from an early age, which we certainly shouldn't have done— and it was also fun jumping out of and into water over the breakwaters, having checked the ground surface of the pools first, wading in the sea (with your hands up a bit to prevent your pony rolling in it with you on board, although it did happen sometimes) and practising flying changes around the pier supports. On a different part of the beach, the sand dunes were great for clambering up and sliding down. Another digression.

THE RIDER'S INFLUENCE IN FAST WORK

The main practical thing we notice about riding fast work is that we need to lean our upper bodies forward somewhat; otherwise, we can find that we are pushed and lurched around and cannot actually sit *with* the horse. A former-jockey friend of mine pointed out that Thoroughbreds galloping at top speed in race conditions can often run at about 40mph/64.4kph (the record being 44mph/70.8kph) and that wind resistance can become a factor at that speed, on a racetrack or elsewhere.

Leaning the upper body forward *from the hip joints*, not the waist, so that the shoulders are above the knees, and keeping the back flat, with shoulders back, chest expanded and upper arms vertical is a good default forward seat for fast work (for rising trot, too, as we have seen). Your stirrup length depends on your leg length and comfort—too long and your balance will be affected, too short and you will be insecure and lose much of the use of your legs for giving cues and helping you stay in position. The fact that some jockeys in flat racing now use *very* short stirrups and barely have their feet in them is entirely up to them! Experiment and find the length that makes you feel secure and in balance at speed, comfortable and effective.

Figure 10.1 Free jumping, starting with very small jumps like this encouraging cross-
            pole, is excellent for introducing a horse, of any age, to jumping, and
            they nearly all seem to enjoy loose jumping. A proper, enclosed jumping
            lane is an asset for training horses and riders in the mind-broadening fun
            of free jumping. This particular horse was very experienced and loved
            his playtime 'pops'.

The main point, as ever, is to learn to keep your upper body as still and controlled
as you possibly can, and to absorb all the movement in your hip, knee and ankle
joints. As your seat will not be actually in the saddle for much of the time, the joints
are where absorption takes place, rather than your waist playing a part as in other
gaits. Keeping your shoulders square with your horse's and looking ahead between
his ears to where you want to go will help you keep still and balanced above the waist
and soft but secure below it. Especially try not to stiffen your ankle joints because
they are key to absorbing your weight and the horse's movements.

Your weight aids or cues in galloping are still very relevant and helpful, and your
horse will be used to them by the time he is being trained or worked in gallop. Your
seat-bones will be out of use, of course, so you will need to rely on dropping weight
down your stirrup leather, flexing your ankle to drop your heel, and letting the weight
on the stirrup bar do the job instead. For a left curve, obviously drop weight down
your left leg *without* leaning over left with your torso which can unbalance your
horse but, if your horse drifts too far left, by all means weight your right leg and foot
instead. The finishing touch is to tap the side of your horse's wither opposite to the
way you wish him to go, for instance, tap the left wither to get him to move right and
weight your right leg.

He may well respond to a standard aid of the rein against his neck. However,
an indirect rein turn might be difficult in gallop because your horse's head will be

pulsing forward and back and your hands, and arms now, will be going with it, so that you can keep as even and communicative a contact as possible. Use your fingers as your first resort, opening and closing them, both hands together, and taking your arms forward and backward only as much as you need to to remain in touch with your horse and to help you keep you both in balance. It always helps to be a minimalist rider.

In the heat of the moment, it is easy to throw caution to the winds and let your upper body swing up and down with the gait, your bottom bang down on your horse's back at every stride, your arms fly around out of position in your understandable attempts to encourage your horse, and your balance out of kilter with the gait. If you can possibly keep your head (and you probably will if your horse is well-trained with reliable responses and has been brought on to not think of galloping as the highlight of his life), keep your heels down and your ankles flexible, your upper body controlled, your joints as absorptive as you can and your arms and hands, fingers first, moving only as much as they have to.

Far from spoiling your fun, you will find that you can actually use your body to help your horse. He gallops in balance without wasting energy swerving around under your unpredictable weight, and so he stays more focused on his job, moving better and more economically, and less likely to sustain injuries.

JUMPING

Some horses, and ponies, are natural jumpers (maybe those forest genes?) and some are not. Some love jumping and some do not. And some can take it or leave it.

Figure 10.2 Lucy letting off steam in the manège albeit from the wrong side of the fence.

Provided a horse at least doesn't mind jumping, they can all be trained to tackle low to medium-sized obstacles, not much over a metre high and with a moderate spread, as part of their general training.

The fast-work forward-style seat or one of its variations is fine for between the jumps and some riders find it easier to adopt it rather than sitting more upright between fences. The main point is to continue to use the classical and ES techniques described in this and other books so that the horse remains clear about the process and can perform confidently and with the least hindrance from his rider.

Any horse will perform better if his need for freedom before, during and after a fence is understood and catered to by his rider. It is clear, when watching any jumping event, that many riders do not do this. The classical and ES way is to approach the fence after an accurate presentation to give the horse a chance, on your in-touch contact, which is your job, and for low to moderate fences to allow the horse to choose his speed (with reservations!) and take-off point.

An extremely important point is to *not* snatch or jab the horse in the mouth on take-off as is commonly done—jumping is certainly an occasion on which the horse must have complete freedom of his head and neck if he is to make a sufficient effort without discomfort and risk of strain injuries. The jab in the mouth, as though the rider needs the reins to keep himself or herself on board, is bad enough but it is compounded by the rider leaning their weight through their hands on the crest or muscles of the horse's neck, blocking the essential stretching out of his neck and head—his balancing pole, remember—which is needed over a jump more than at any other time apart from when getting up from the ground.

JUMPING SEQUENCE

INTRODUCTION

These diagrams were drawn from a photographic sequence of a horse jumping loose/free. The black, horizontal line represents the top rail of a fence in the photographs. This sequence shows the natural biomechanical functioning of a jumping horse, which we should always try to emulate. The head remains fairly level with the rail throughout the jump although the body and leg positions change considerably. In effect, he 'jumps around his head', which is the centrepoint of his jumping balance. Therefore, the rider must allow the freedom of the head and neck. If the rider interferes by hauling his or her body forward and up via the bit, or to maintain security, the jump will be very difficult for the horse, less successful and potentially injurious to his mind and body. (*Refer to drawings on pages 79 and 188, for comparison.*)

'*Horses use their heads to balance. If it can't balance, it can't use its stride, run fast or jump. Any horse does its best to stay on four feet but it needs to use its head.*' Leading equine physiotherapist, the late Mary Bromiley, interviewed in *Horse & Hound*, 16th May, 2019.

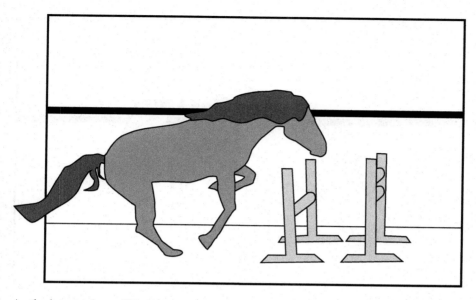

As the horse takes off, he starts to stretch out his head and neck.

The horse pushes off with his hind legs, folds up his forelegs and stretches his head and neck forward and out.

As he leaves the ground, his head and neck are fully stretched out, allowing his body to follow and work naturally with the least effort.

In full flight, the head and neck are still stretched right out, emphasising the importance of freedom when jumping.

In the descent, the head and neck, still stretched out, begin to adjust to the horse's change of balance.

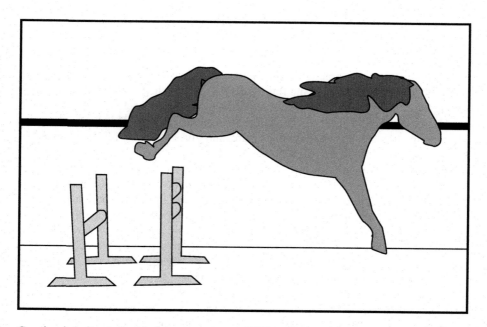

Coming in to land, the hindquarters are now higher than the head, which has remained level with the fence rail throughout.

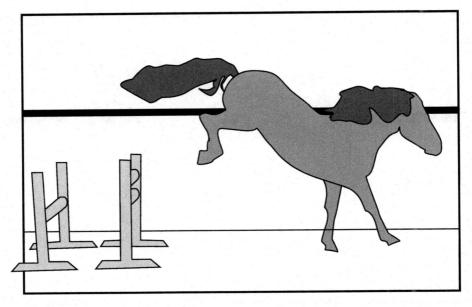

There is tremendous stretch down the back of the landing foreleg as it hits the ground. The hindquarters are well above the height of the head, so the rider needs to stay just out of the saddle to avoid banging the horse on his back. The head is below the fence rail for the first time since take-off, as the horse appears to raise his head and neck to adjust his balance to the thrust of landing.

THE RIDER'S INFLUENCE WHEN JUMPING

The bad effects of this practice can be lessened if the rider learns to slip the reins, that is, to let them run through the fingers during the stretch, gathering them up again during the descent and landing. However, the very best way to actually help the horse over an obstacle is to use the classically based jumping seat. This keeps the rider closer to the horse, is stiller and more secure, and gives the horse full freedom of his head, neck and body, all in a perfect natural balance of horse and rider.

The classical jumping seat was used internationally by most of the world's best jumping riders, and down through the ranks of more ordinary mortals, being taught as standard at most good riding schools, from the end of the nineteenth century. This turning point in jumping riders' technique came about because of the development by Italian cavalry officer Federico Caprilli of his 'forward seat'. Photos of Italian cavalry began to appear of riders and horses using this seat over fences but also cross-country hazards—like near vertical banks with multiple riders leaning forward sliding down them on their horses, which were enough to put the fear of death into more timorous souls.

Slight variations on the forward seat were used, but the classical-style Caprilli seat was widely used internationally until about the 1960s. Then we started to see more and more riders adopting a seat further away from their horse's body, balancing themselves with their hands, holding too-short reins and so fixing the horse's head

Figure 10.3 The light seat. There are variations of this light seat, but this is an impression of its core principles. The rider's seat brushes or is just out of the saddle. His or her balance is directly over the horse's centre of balance and is felt by him in the area under the stirrup bars, on his back. These usually take all the weight of the rider, transferring it on to a small area on each side of the spine just behind the withers, emphasising the need for good saddle fit and weight distribution to protect the back. It is a great help if the rider learns to take some weight down the insides of the thighs and knees and adds to the security of this seat. The straight-line guide here is shoulder > elbow > knee > ankle joint or front of the rider's foot. The lower leg is used further forward when riding at faster gaits downhill, which increases the pressure on the horse's back.

and destroying his efforts to balance himself and make his jumping effort. Totally counterproductive. They would throw themselves too far forward, out of balance, up their horses' necks, plus returning to the saddle too soon during the landing and getaway and so banging the horse on the back with their seat and jabbing the horse in the mouth—altogether extremely poor reward for a horse trying to do his best.

The older, better balanced and therefore safer seat, with its accord with the horse's movements, is as follows:

- During the rider's own training, trouble is taken to learn an independent, balanced seat so that reliance on the reins and supporting themselves on the crest of the horse's neck are not needed.
- During take-off, the rider folds his upper body *down* from the hip joints so his chest is close to the horse's withers or the lower part of his crest, at the same time pushing his seat *back and down* towards the cantle. This puts the rider close to the horse and in non-interfering, secure balance with him.

Figure 10.4 A very different picture from the conventional, modern jumping position. Here, the rider is allowing the horse's biomechanics to function naturally. The rider's balance means he or she is safe and secure, folding the torso down from the hip joints, not the waist, close to the horse's withers and neck. The back is kept flat and the shoulders back and down, the seat is pushed *back* towards the cantle to enable the fold and secure the body position. The lower legs remain more or less vertical and *down*, with the heel pushed down as a secure basis for the whole seat. Crucially, the hands follow the horse's mouth *forward and down* to give him essential freedom to stretch fully out over his obstacle without having to lurch over it uncomfortably due to a restricted head and neck, with the rider's hands fixed on the crest of the neck. This must surely create pain in the mouth and probably strain injuries to the neck, shoulders, back, hindquarters and hindlegs due to the unnatural stresses on them in the more restrictive, and now almost universal modern technique.

- The rider's arms and hands are moved *down and forward* towards the horse's mouth as he stretches in flight over the fence; ideally, the rider lets the horse's mouth take his hands, known as 'following the mouth' or 'going with the mouth'.
- On the descent and landing, the rider keeps his seat *out* of the saddle while raising his torso up again, and brings his hands back as the horse's head rises to normal again, with a very light contact.
- After the getaway, the rider can return his seat to the saddle if he is using a more upright seat between fences; otherwise, a forward, light seat is used.

This seat is suitable for most jumping genres, including racing over fences, the exception being very high and *puissance* fences, when it is quite understandable that the rider will need to align with the horse's more upright body position on take-off. However, it is even more important that, in these situations, the horse does not receive a jab in the mouth at any time during his jumping effort. The most usual time for this to happen is on take-off as the rider uses the reins, that is, the horse's mouth, to raise his upper body up the horse's neck and, to add insult to injury, often props on the horse's crest with the hands on a short rein, pulling in the horse's head and making life really difficult and quite probably painful for him.

If a rider really needs something to hold on to during riding, a neck-strap can easily be fitted. For something further forward, as in high jumps just described, there is nothing wrong with *not* plaiting up a horse's mane but leaving it free for the rider to hold, if needed. Plaits are often put in too tight, anyway, and kept in for too long, and can cause certainly discomfort but possibly pain, so can be counterproductive and a welfare issue. Many people, including your author, much prefer a well-kept *un*plaited mane and I am sure the horses do, too. Fashion and appearance should certainly come second to the horse's welfare.

Teaching and using the previously mentioned techniques, life would surely be so much better for jumping horses in all sports, competitive and otherwise. From the rider's viewpoint, the jumping seat described earlier gives the rider a much better feel because of its increased stillness, closeness and security, and, from that of the horse, the rider's weight is closer to his body and therefore not so top heavy and is easier to manage because the rider does not move around and disturb the balance of both parties as often happens in the more usual, modern seat. The sensation for both parties is much more exhilarating and, because the horse can use his body more naturally, less potentially injurious.

In show-jumping, horses are often required to make a steeper parabola (rounded arc) over their fence so it is even more important that their heads and necks are not restricted, yet so often the priority seems to be for the rider to stay on board despite possible discomfort or, possibly, pain for the horse jumping in a restricted way. For cross-country, the seat described was very often, indeed normally, used because of its 'neatness' and efficiency over fences that were often much more frightening for the rider than today's—which would certainly terrify me. In the past, the horse's mouth was surely regarded as more sacrosanct than it is today. The rider's lower legs across country often need to be brought forward over drop fences, banks and so on, but the close upper body and freedom of the head can still be adopted, for the benefit of the horse.

For interest, the old-old jumping seat taught and used in some quarters, particularly the hunting field where it is still sometimes seen, involved the rider leaning forward on take-off over a fence but *backward*, slipping the reins considerably, during the descent and landing, necessitating a thrust forward again as the horse got away. Obviously, the rider's body is still swinging around in this seat and the backward thrust in the descent, plus the lurch forward again on landing, really played havoc with the horse's balance, control and ability to function effectively. The backward swing, Caprilli found, seriously hampered the action of the hindquarters during the descent, putting the horse at a considerable disadvantage in the landing and getaway, and if the rider did not slip the reins enough, the horse's mouth received a major and painful bang in the mouth. Enough said!

Figure 10.5 No one can doubt the bravery of cross-country riders, and most of us can only envy it, but the very nature of the discipline makes perfection even less attainable than in other equestrian disciplines. Here, the rider is having to keep position in the saddle by using the reins and the horse's mouth to stay on and in balance. This cannot help the horse's efforts to jump out of the water as he needs the freedom of his head and neck for a best effort, and the pressure in his mouth must be considerable.

The modern jumping seat often involves riders bringing their bodies too far forward up the horse's neck. This positioning does not accord with the horse's own balance, as described, but interferes with it. The rider's lower legs, in such an effort, are usually swung up and back, toes down, and provide no security at all. In addition, the hands, pressing down on the crest of the horse's neck and leant on by the rider, are nearly always firmly fixed so that much of the rider's own weight is concentrated on the horse's neck, which needs to be free for balancing. If, at the same time, the reins are too short, preventing the horse's head and neck stretching out to enable him to land smoothly and safely, the horse's efforts are being very significantly hampered by the rider in several ways.

RACING

Racing on racecourses, whether on the flat or over hurdles or fences, is often seen as a world apart from the 'ordinary' horse world, but horses will always be horses and their minds and bodies function just the same. Of course, the very short stirrups used in racing, even 'chasing nowadays, make a big difference to the jockey's seat.

As I have never trained racehorses myself or ridden in a race, only in work, I can only give a mainly theoretically based opinion, but it does seem to me that, if some jockeys would sit stiller and modify the exaggerated movements of their arms and hands, and put paid to the consequently flapping reins, this would be less distracting to their horses and enable the latter to balance and move better. It must also improve the jockey's balance, which would be all to the good.

When watching racing, I have noticed that, often, the usually amateur jockeys who compete in other equestrian disciplines as well as racing have much stiller seats, seem physically more 'together', don't wave their arms and hands around as though hailing a taxi and completely 'dropping' their horses, don't use their whips as much and stay more in light contact with their horses' heads while giving them plenty of freedom in their extended gallops, and they have their share of wins.

The issue of jockeys banging down on their horses' backs has been dealt with earlier, but it is one more occurrence in racing that goes against the horse's chances of doing his best. The topic of tongue-ties has also been mentioned and they are banned in some countries. My feeling is that if a horse cannot race without his tongue and, I understand, the full functioning of his throat area, being artificially restricted, he's been put into the wrong job.

ENDURANCE RIDING/RACING

This sport, at the time of writing, regularly receives very adverse publicity in the equestrian press and, it seems, with good reason. From the standpoint of this book, I want to mention that the usual way of riding of endurance riders, that is, with loose reins most of the time, is doing their horses no favours.

The argument is usually that their horses do not need to be on the bit or to have any kind of dressage-type strengthening and lightening because their discipline does not require it. I beg to differ absolutely. The stronger backs and hindquarters the horses would develop would enable them to work over very long distances, and at speed, without the potential considerable backache, stress and muscle strain many of them probably suffer from, according to an endurance veterinary surgeon I asked about this. The lightening of the forehand, too, would take so much weight and stress off the forelegs.

I am not, obviously, saying that a horse has to travel along in a competition on the bit for most of the time, but if he had regular training to enable him to do so, as described in detail in this book, it would make him stronger and better balanced, and so able to work at the distances required with less stress.

The riders, too, can help to relieve their horses' backs by not standing in the stirrups in the belief that this eases them. Again, as explained earlier, it does not do this, but concentrates the weight just under the stirrup bars, so potentially making things worse for the horse. By using a light seat instead and taking some weight down the insides of their thighs and knees, they can spread it and ease the horse's burden that way.

IN A NUTSHELL

It might seem strange that classical riding could be applicable to faster work, modern show-jumping and cross-country riding, racing and also endurance racing and riding,

Figure 10.6 It's good to lead your horse around in hand after work, with a loosened girth and run-up stirrups. It cools off and relaxes the horse and gives the rider something to think about. We hope the horse enjoyed his work, whether schooling or a hack, and that the reader of this book puts it down with plenty of horse-friendly ideas to consider for the future. Thank you for taking the trouble to read it.

but that is the case. The ES principles, of course, apply to them as well, and I hope that this chapter will show the advantages of these two ways of riding in fast work, increasing safety and horse welfare.

The faster gaits are all considered with some variations in relation to fast work and jumping, plus gait abnormalities. Adapting the classical seat for faster gaits is described, plus the application of weight aids, which are still very relevant. The classical, forward-based jumping seat is described and explained, together with the modern and old jumping seats, with their disadvantages. The important points in using any seat are to make sure the horse has real freedom of his head and neck to make his effort and, particularly, that he does not receive a painful bang in the mouth from an unbalanced rider during the process.

The racing seat is also considered and, respectfully, some improvements suggested. Endurance racing and riding do not escape, either, and the advantages of schooling endurance horses to be able to go on the bit during training are explained, plus the disadvantage of their riders standing in the stirrups in the mistaken belief that it relieves the horses' backs. In fact, it concentrates the pressure, so riders could, with advantage to their horses, adopt a light seat instead and learn to take some weight down the insides of their thighs and knees.

Conclusion

I assume that if you have read this far, your interest has been stirred in the topics covered in this book. I also hope that if you are or plan to be a teacher, you will want to explore them further (*see* 'Recommended Reading' and 'For Your Information' next) and pass them on to your students and clients.

If you do not teach but are involved with horses professionally or as an amateur, I hope you will consider the maybe rather different way of dealing with them presented here and also that Chapter 6 will encourage you to think in a new, open way about every aspect of the horses that depend on you.

When we are presented with new, different ideas, or older, unfamiliar ones, we can feel threatened because they can disrupt our thinking about something that may have become comfortable and second nature to us. My purpose in writing this book was, indeed, to try to persuade people to think differently from what has become normal with an open mind, a fair, compassionate attitude and the welfare and well-being of horses foremost in their minds. You will find in these pages a blend of the best of the old with the best of the new from which both horses and humans can benefit.

Three generations of equestrians have grown up without the understanding of equines taken for granted by their parents and grandparents. Society has largely lost the feeling for and knowledge of the horses and ponies that were once a major part of it. We witness wonderful improvements in veterinary science and related modalities, yet riding and training have, in my experience and opinion, deteriorated considerably. Many practices and beliefs have developed that are not at all appropriate for horses and, because there are fewer people left to pass on older, better ways, attitudes have taken root that many people do not realise are not at all horse-friendly.

On the other hand, scientific research is uncovering all the time now new knowledge about how our horses really function, physically and mentally. In some cases, older practices and knowledge regarding horse care and training have been confirmed to be totally appropriate.

The only good way forward is to truly put the welfare and well-being of our horses and ponies first, in practice. From that vantage point, I wish us all, equine and human, a happy future.

Susan McBane

DOI: 10.1201/9781003121190-11

Recommended Reading

The books here, some with DVDs and websites, are listed by author in alphabetical order. Some are specifically about classical riding or Equitation Science and others aren't, but they promote the ethos of a humane, effective, 'thinking' attitude to horses. They can all be located by searching the Internet. *However, for the sake of the future of the book publishing industry, please order your books from independent book retailers and dealers whenever possible.*

BARBIER, Dominique and CONROD, Liz
BROKEN OR BEAUTIFUL: The Struggle of Modern Dressage (www.xenophon-press.myshopify.com). Just published at the time of my writing *Fine Riding*, this is one of the most important books on equestrianism ever to be produced and is essential guidance and reading for anyone genuinely concerned about equine welfare and well-being. It exposes fully the potential dangers of some modern riding techniques and attitudes and makes disturbing reading in some parts, but there is plenty of light at the end of the tunnel for those willing to see it. I should love this book's message to be read, taken to heart and acted upon by everyone with an interest in horses, for the sake of equine welfare, well-being and prosperity.

CAPRILLI, Federico
The Caprilli Papers, Principles of Outdoor Equitation, by Captain Federico Caprilli, translated and edited by Major Piero Santini. The forward jumping seat and its evolution explained, plus much more. Now rare and, for most people, not cheap, it is nevertheless totally absorbing and an enlightening read.

HEUSCHMANN, Gerd
Tug of War, Classical Versus 'Modern' Dressage: Why Classical Training Works and How Incorrect 'Modern' Training Negatively Affects Horses' Health, with DVD, the first of the author's books to shake up the modern, conventional horse world. Get the second edition. His other three books, so far, are *Balancing Act; Collection or Contortion?;* and *Classical Schooling with the Horse in Mind.* Priceless!

KILEY-WORTHINGTON, Marthe
The Behaviour of Horses in Relation to Management and Training; Equine Welfare; Horse Watch: What it is to be Equine and *Animals in Circuses.* More books to challenge conventional thinking from a scientist and horsewoman never afraid to speak her own mind.

LOCH, Sylvia.

The founder of the international Classical Riding Club, the author, now semi-retired, is one of the best-known practitioners and teachers of classical riding today. My favourites of her several invaluable and always educational and entertaining books are *The Classical Rider, Invisible Riding, The Balanced Horse* and *The Rider's Balance*. Cornerstone books for your library.

McBANE, Susan

Please forgive me for listing a few of my own books. *Fine Riding* is my first book to cover Equitation Science. My two books which readers seem to find particularly helpful for classicism and which complement each other are *Revolutionize Your Riding* (which concentrates mainly on the horse) and *Horse-Friendly Riding* (which concentrates mainly on the rider). *Bodywork for Horses* gives DIY techniques for giving your horse a massage, a Shiatsu treatment, correct stretches and more.

McGREEVY, Paul

Equine Behavior, 2nd edition. A very comprehensive, scientific book accessible to lay readers, by one of the key instigators and practitioners of Equitation Science worldwide. Extremely thorough—and you'll feel you've got a PhD when you've finished it.

McGREEVY, Paul, WINTHER CHRISTENSEN, Janne, KÖNIG von BORSTEL, Uta and McLEAN, Andrew

The first edition of *Equitation Science* quickly became an indispensable volume for any Equitation Science student, horseman or horsewoman wanting to acquire a full and thorough knowledge of this new equestrian science. This second edition, by four authors, has been fully revised with new contributions covering all the latest developments in ES up to, obviously, the time of writing. It is totally necessary for anyone who wants to really get to grips with ES for either or both academic and practical purposes. Non-scientists won't find it an easy read, but you can always refer to the excellent index and glossary for further helpful references and down-to-earth help. If you seriously want to understand ES, I can seriously advise you to buy this book!

McLEAN, Andrew and Manuela of Equitation Science International

Academic Horse Training: Equitation Science in Practice, a sizeable hands-on how-to book on the principles of ES and how to apply it, in detail. Essential. Available from Equitation Science International at www.esi-education.com/shop/ plus several other key books and DVDs on ES, straight from the horse's mouth, as it were. Browse away!

PODHAJSKY, Alois

Complete Training of Horse and Rider in the Principles of Classical Horsemanship is still the classicists' bible decades after its first publication, by the former head and Director of the Spanish Riding School. Pure classical principles from initial handling to airs above the ground, plus high regard for the horse. A must. Search his other books. *The Riding Teacher* is particularly applicable in view of the modern changes in horsemanship education.

ROBERTS, Tom

No equestrian library is complete without Tom Roberts's books. Wonderful horse sense from this premier Australian horseman, as in *Horse Control and the Bit, Horse Control—The Rider, Horse Control—The Young Horse* and *Horse Control Reminiscences*, also *Go Forward, Dear*, a biography of Tom Roberts by Andrew McLean and Nicki Stuart.

ROLMANIS, Debbie

Biomechanics for the Equestrian: Move Well to Ride Well. The better you move and keep your body in good working order, the better you will ride and the better your horse will go. This book is more about how you move all the time rather than a schedule of punishing exercises—there are some in this book, and they aren't punishing. Accessible and definitely helpful in practice, in my own experience.

SCOFIELD, Rose M.

Many of the behavioural problems people experience with their horses are due to confusion in the horse's mind and are not *bad* behaviour but self-defence. Also, it seems that many more horses and ponies are euthanised today than in the past due to perceived dangerous behaviour. *Solving Equine Behaviour Problems: An Equitation Science Approach* will be of immense help to owners who just don't know what to do next. It is a most valuable and groundbreaking book, easy to read and understand, to have by you or consult in times of trouble.

SKIPPER, Lesley

Lesley is a friend and colleague on *Tracking-up* magazine (*see* 'For Your Information' next). Her several books are always highly relevant and to the point, my favourite being *Let Horses Be Horses: The Horse Owner's Guide to Ethical Training and Management*, which should be required reading for examinations. Search her other books for sound common horse sense and nononsense language.

STANIER, Sylvia

Sylvia was a valued friend and an old-school classicist with very varied equestrian interests and abilities. Do search her life and career on the Internet. Her approach to horses was to always find out what each individual horse was like, then treat him or her accordingly from her vast store of knowledge, skills and compassionate, no-nonsense, common horse sense. I recommend three of her books in particular: the set of two little hardback books *The Art of Lungeing* and *The Art of Long Reining*, plus her invaluable *The Art of Schooling for Dressage*. Essential reading for modern riders looking for, or willing to consider, something better. Most of my readers would also find *Classical Circus Equitation* a real eye-opener.

WILSON, Anne

Anne is also a friend and part of the team on *Tracking-up* magazine. Her book *Riding Revelations: Classical Training from the Beginning* is a very detailed but simply expressed book to reassure anyone rightly dissatisfied with much of the modern approach to horsemanship. Photographs and diagrams with clear instructions in their captions enhance the learning process. A really encouraging read that lets you know that You Can Do It!

For Your Information

Sources of information, education and interest mostly relating to the subject matter of *FINE RIDING*, presented in alphabetical order

ASSOCIATION OF PET BEHAVIOUR COUNSELLORS

The APBC is an international network of experienced and qualified pet behaviour counsellors who work on referral from veterinary surgeons to treat behavioural problems in horses and other animals. APBC members are able to offer the time and expertise necessary to investigate the causes of unwanted behaviour in animals and outline practical treatment plans that are suitable for their clients' circumstances. For further information, visit *www.apbc.org.uk.*

BLACK CAT PHOTOGRAPHY

If you are searching for a professional, independent library of equestrian photographs, Black Cat Photography has hundreds of them, many of which appear in this book. Whether you want one or many, you could find what you want here. Commissions also undertaken. For further information, visit *www.black-cat@virginmedia. com.*

CLASSICAL RIDING CLUB

The international Classical Riding Club, founded by probably Britain's premier classical rider and teacher, Sylvia Loch, is no longer active but its website is still available, with its invaluable archive of many hundreds of articles freely accessible to all. It also features the CRC Trainers Directory listing classical trainers all over the world by name and by location. Find out more at *www.classicalriding.co.uk.*

EQUITATION SCIENCE INTERNATIONAL

Equitation Science International is the education wing of the Australian Equine Behaviour Centre, founded and run by Andrew McLean and his wife Manuela. In my experience, their staff, also, are most efficient and helpful. ESI's stated mission is 'to educate horse riders and handlers in equitation science to enable efficiency and safety in all horse interactions'. They offer various educational opportunities online and 'live', including the accredited and highly regarded Diploma of Equitation Science, a horse training and coaching qualification that provides a full understanding of horse training and is appropriate for all equestrian disciplines and activities. Enjoy browsing their site, shop and practitioner list at *www.esi-education.com*.

HUMAN BEHAVIOUR CHANGE FOR ANIMALS

As any teacher knows, the most difficult part of teaching people in relation to their animals is to get them to change their behaviour towards them and to use more effective methods in their dealings with them. HBCA specialises in this skill. It operates internationally, runs a consultancy, is involved in research and organises training sessions and events both live and online. Visit its website at *www.hbcforanimals.com*.

INTERNATIONAL SOCIETY FOR EQUITATION SCIENCE

The ISES is a non-profit organisation that 'chiefly aims to facilitate research into the training of horses to enhance horse welfare and improve the horse/rider relationship'. It is a membership oganisation, with various membership categories open to people with appropriate qualifications and experience. However, non-members are also welcome to attend its annual conference, with often ground-breaking presentations, held in a different country each year. ISES sells its wedge-shaped taper gauge to individuals and organisations, to be used for gauging the appropriate fit, as regards tightness, of horses' nosebands: the gauge also measures the circumference of bit mouthpieces, to check compliance with FEI rules. It issues periodic position statements on matters of importance in equine training, management and welfare and is one of our most important equestrian organisations. Its website, which I regard as essential reading for all equestrians, is at *www.equitationscience.com*.

THE SADDLE RESEARCH TRUST

The Saddle Research Trust is a charity registered in England. It aims to 'promote the health, welfare, safety and performance of ridden horses and their riders'. It has a strong educational ethos, with freely available, open-access, web-based resources and enewsletters—no membership is needed. Its regular public conferences present essential developments in equestrian topics related to the health and welfare of horses and riders. Find out more about their crucial work at *www.saddleresearchtrust.com*.

TRACKING-UP MAGAZINE

Tracking-up is a non-profit, voluntary, quarterly, equestrian magazine, available internationally, which concentrates on true classical riding, Equitation Science, and ethical horse-care and management methods. It is available by subscription or single-copy purchase, in hard copy or digitally. To maintain its independent view and to enable it to say what needs saying on behalf of horses, it has no commercial or sponsorship connections. Its Editor is the author of this book, Susan McBane. If you drop her an email at horses@susanmcbane.com, she will send you full details of how to acquire your copies. You will be most welcome as a *Tracking-up* reader.

Index

Page numbers in *italics* indicate figures.

abducting 91
adducting 91
advanced balance 134, 155
aids (signals/cues)
    in canter 149, 169–170
    classical buttons 149
    classical conditioning and 89, 144–145
    classical direct turn 147
    classical seat and 145
    confusion and 73, 77, 86, 88
    consistency and 99
    correct use of 10, 72, 90–91, 144–145
    equine mental health and 45
    gaits and 91–92
    indirect rein of opposition 147
    indirect turn 147
    individually identical 90
    leg and seat 148–149
    light 6, 74, 77, 81, 121, 152, 155, 170
    in modern riding 45, 73, 90–91
    operant conditioning and 88
    opposing 45, 73, 90
    predictability and 64
    reins and 145–148
    releasing/stopping 75, 86, 105–106, 140
    riding curves 174
    riding with one hand 146–147
    seat and weight 172–173
    for stopping 21
    touch and 24, 27
    in trot 149, 168
    using less hand 173–174
    vocal sounds 23–24, 62–63
    in walk 149
    weight aids 6, *73*, 93–94, 172–174
animal welfare 6, 68, *see also* Five Domains of
        Animal Welfare
approach conditioning 87
associative memory 84

backing 99, 102–104, 117
balance
    advanced 134, 155
    'on the bit' 121, 128, 130–134

classical riding and 6, 10
development of physique and 127
on the forehand 132–134, 155
forward *156*
horizontal 121, 130, 133–134, 155, *156*, 160
of horse and rider 34, 130–131, *133*, 173
jumping and 17
lungeing and 108
modern dressage and *73*
modern riding and 77
rearward *156*
riding from back to front 129–130
saddle and 131–132
self-carriage and 93–94, 106
three-quarter gallop and 92
barefoot working 65
behaviour
    basic needs and 82
    confusion and 112
    differential reinforcement and 87
    discomfort and 11, 34, 77–78
    equine mental health and 45, 75–77
    Five Freedoms and 51
    horse well-being and 51
    human judgements on 35, 37
    incorrect operant conditioning and 88
    lack of agency and 49
    modern riding and 160, 179
    negative reinforcement and 98
    rewards and 25
    self-defensive 11, 19, 44, 82, 112, 179
    showy action and 78
    weaving *56*
behind the vertical (BTV) posture 33, 77, 135
bend 135
biomechanics 12, 71, 81, 121, 127
bit contact
    ancient classical riding and 1
    'on the bit' 121–122
    faulty posture and 128–129
    levels of 123–125, 128
    light 3, 6, 32, 66, 98–99, 125, 127
    loose reins 124, 128
    problematic techniques in 8, 19, 77, 127